BUSINESS ISSUES, COMPETITION AND ENTREPRENEURSHIP

INNOVATION AND GLOBAL COMPETITION

THE CASE OF KOREA

Business Issues, Competition and Entrepreneurship

Additional books in this series can be found on Nova's website under the Series tab.

Additional e-books in this series can be found on Nova's website under the e-book tab.

BUSINESS ISSUES, COMPETITION AND ENTREPRENEURSHIP

INNOVATION AND GLOBAL COMPETITION

THE CASE OF KOREA

FREDERICK BETZ
YOUNGRAK CHOI
WHASIK MIN
WON KOO PARK
AND
DONG WOO SHIN

nova publishers
New York

For permission to use material from this book please contact us:
Telephone 631-231-7269; Fax 631-231-8175
Web Site: http://www.novapublishers.com

Library of Congress Cataloging-in-Publication Data

Innovation and global competition : the case of Korea / editors, Frederick Betz, Youngrak Choi, Whasik Min, Won Koo Park and Dong Woo Shin.
pages cm
Includes index.
ISBN 978-1-62808-713-0 (hardcover)
1. Technological innovations--Economic aspects--Korea (South) 2. Economic development--Korea (South) 3. Industries--Korea (South) 4. Competition--Korea (South) 5. Korea (South)--Economic policy. I. Betz, Frederick, 1937-
HC470.T4I55 2014
338'.064095195--dc23

2013048021

Published by Nova Science Publishers, Inc. † New York

CONTENTS

PREFACE

We will analyze the economic development of South Korea within the basic sociological dichotomy of *Individual and Society*. As we know, society is composed of individuals, and individuals act within the context of society. Economics consists of agents acting within the structure of a market. The economic agents are individuals, and markets provide the social structure in which their actions occur. One can vividly see this interaction in the modern history of Korea. Much of the South Korean national success has depended upon key individuals, individuals who displayed outstanding abilities in business and in government. Together, they built a Korean structure of an industrialized export-oriented economy within a democratic society – leadership and structure.

From this analysis, we will see evidence for a basic economic proposition. Economic development is not simply a matter of social structure (Laissez-faire free-market), nor is it simply a matter of individual leadership (visionary leaders). Economic development requires both proper structure and good leadership. However, good leadership and proper societal structure together is not a simple historical process, nor is it even historically certain.

Bad leadership can corrupt societal structure; and bad societal structure can promote corrupt leadership. One can see this in comparing North Korea with South Korea. The history of North Korea is that of an unjust political system and an inefficient economic system, built under bad leadership. The dictatorship of Kim Il-sung created a communist society that was inefficient and ineffectual, neither feeding nor freeing its people. Structurally under the leadership of Kim's son and grandson, the dictatorship continued as a 'miliary-first policy'. In contrast, the history of South Korea is one of growth to an efficient and globally competitive economic system, evolving at the same time a democratic political system. In the South Korean story, entrepreneurs competed and built businesses to operate in a self-organizing capitalistic economy. Politicians compromised but did not tolerate corruption and formed a government to operate in a self-organizing democracy.

The societal concept of a 'self-organizing society' is a revealing way to understand how South Korean economic and political systems operate. Although there has been strong leadership in industry and in government, there has been no single person in charge of the whole society, in contrast to the North Korean dictatorship.

No one is in charge of a self-organizing society. While a single business person (entrepreneur) can control a business system, no single entrepreneur is in control of the entire economic system. While a government official, such as a president or prime minister, can control the executive branch of government administration, no single politician is in control of the entire political system. This is why any capitalist democracy is self-organizing – not controlled by any central authority, business or government. Then how can the socio-technical systems of self-organization work? How can the economic system work? How can the political system work? This has been the central problem of traditional economic theory; and this has been the central problem of traditional political theory.

One answer we will see that stands out, in the history of South Korea, is this. 'Laissez faire' economic ideology (which has emphasized economics but demeaned politics) has been incomplete for effective economic development and democratic growth in an industrializing society. Certainly a free-market economy is necessary for economic growth. Yet there was a proper role of government in economic development. Government needs to construct a proper context for economic activity – including an effective societal infrastructure, effective regulation of free markets, and honest rule. When the political system is not just, the economic system can collapse under rebellions and civil war; when the economic system is not efficient, then the political system can collapse under riots and dictatorship.

Thus critical to the South Korean experience has been the development of both a free political system and a free economic system – in proper cooperation. We will analyze the South Korean partnership between industry and government. What is striking in this analysis was the delicate balance between patriotic political leadership and competitive business leadership. It was the balance that ensured both (1) a capitalistic society, creating prosperity spread widely over both capitalists and labor, and (2) a democratic society, which elected governments ruling for the public good rather than only for private interests.

Good-leadership-in-a-good-structure is a balance, but not an easy one. This generalization, which we will see illustrated in the history of modern Korea, is a basic principle, fundamental for any national economic development. In a self-organizing democratic-capitalist nation, industry and government need to properly cooperate for prosperity and justice. That partnership must be very ethical: with business people not corrupting politicians and politicians not formulating laws for the benefit only of the few.

Frederick Betz
Youngrak Choi
Whasik Min
Won Koo Park
Dong Woo Shin

Korea University and Gyeongsang National University and Nano Corporation
October 2013
Seoul, Korea

Chapter 1

KOREA AND GLOBAL CIVILIZATION

INTRODUCTION

'Industrialization' is a necessity but also a challenge for modern nations. Industrialization is the application of science and technology to the means of producing the goods and services needed in society. Science began only in the late 1600s, and industrialization began only in the late 1700s. The modern world has been experiencing the industrialization of nations for three hundred years, first in Europe, then in America, and next in Asia. Here is the current situation.

The world is continuing to be industrialized, but now within a global context. The point of industrialization is to attain a prosperous and stable society. Prosperity means that a nation's population all can live above a poverty level. Instabilities in a society include financial crises, economic depressions, foreign wars, insurrections and civil wars. As yet no industrial nation has achieved both prosperity and stability.

Industrial prosperity first occurred in Europe in the nineteenth century (beginning with the industrial revolution in England in the late eighteenth century). But in the beginning of the twentieth century, all culminated in the major instability of World War I. Industrial prosperity next occurred in the United States in the beginning of the twentieth century, but was temporarily stopped by the Great Depression of the 1930s. During and after that First World War in Europe, prosperity was not attained in European nations and several countries turned to communist and fascist dictatorships (Russian communist revolution in 1917, Italian Fascist take-over in 1922, and German Nazi coup in 1933). All this culminated in World War II.

After that Second World War, a Cold War followed between capitalistic democracies and communist dictatorships – that culminated in the collapse of communist governments. It was during and after the Cold War, that the industrialization proceeded in all the Asian nations. Yet in the beginning of the twenty-first century, global prosperity and peace had not been attained. Particularly, the financial system of the global economy was unstable (with the 1997-2000 Asian financial crisis, the 2007-08 global financial crisis, and the 2010-14 Euro-dollar crisis in Europe). Thus historically we are not yet certain how an industrialized nation can be both prosperous and stable.

Instability impacts the economies not only of industrially developed countries but also developing countries. The Asian Financial Crisis of 1997-99 impacted the economic development of Thailand, Indonesia, South Korea, and several other countries. The big picture of a nation's economic system is studied in a topic called macro-economics. But while 'big pictures' are always useful, often it is the 'little pictures'—microeconomics – in which the hard 'graininess of reality' is seen and felt. By examining individual cases, one gets a sense of the harsh impact which instability can have upon a society.

DAEWOO'S BANKRUPTCY

For this reason we will begin our study of industrialization in a global world by looking at two bankruptcies of major global corporations when the twenty-first century began. These were: first the1999 bankruptcy of the Daewoo Group in South Korea and secondly the 2009 bankruptcy of General Motors in the United States (GM, ironically, had earlier acquired the bankrupt Daewoo Motor Company). In the bankruptcy of global corporations, one can see clearly the complexity of global factors in play. It is in the international financial crises that the soundness of global corporations are tested, as well as the soundness of national economies.

In that Asian Financial Crisis (which we will examine in depth in later chapters 6 and 7), the international financial community attacked, successively, the monetary exchange rates of Thailand, Indonesia and South Korea. Consequently in each country, foreign loans to local banks were recalled, because of the loss of loan profits due to falling rates. Credit in each country froze; businesses collapsed and unemployment surged. Each nation was plunged into a depression and a government crisis. Yet South Korea was the country which recovered most quickly from the crisis and went on to become a major competitor in the world. Later we will review how and why South Korea successfully and quickly recovered.

Yet in any national crisis and recovery, some individuals and groups are not successful. Daewoo was one group caught in the crisis. Donald Kirk wrote: "Lee Jeong Seung, a Daewoo spokesman, said: 'Daewoo Motor . . . planned to enter the American market in 1999 as part of a group plan to increase overall exports from $15 billion in 1998 to $17 billion in 1999.' Lee cited three reasons why he believed Korea now faced economic difficulties – 'short-term foreign exchange, non-competitiveness, and the trade deficit.' In response, he said, 'we have to export more.'" (Kirk, 1999)

In 1999, why did this strategy of 'more export' not work for Daewoo Motors? It could not and did not because of the collapse of the credit market in Korea. To produce and sell automobiles, one first needs credit to pay manufacturing costs. After production, cars are sold to recover costs (depreciation, supplies, labor) and make a profit. From the time of production to the time of sales, any business needs credit; and banks provide credit. When banks fail, credit disappears and production ceases. This is the connection of banks to business recessions – the freezing of credit which halts economic production. This was the bind in which Daewoo found itself in 1999. It was a financial problem, not an export problem. The financial bind asphyxiated Daewoo.

Daewoo was one of the largest of the many chaebols which together built the economic miracle of South Korea. At the time of its demise, Yonhap News reported: "The story of the creation of Daewoo was also the story of the rise of South Korea. Over just three decades, Kim Woo Chong, a former shipyard worker of modest means, built, through energy, guts and determination, a global industrial giant that reached from automobiles and electronics to financial services and construction. The conglomerate's allegorical power has therefore given its decline an uncomfortable significance for South Koreans. On August 16th, Daewoo's long-suffering creditor banks announced a government-orchestrated plan to dismantle the sprawling group, signaling what they hope will be the final chapter in a protracted crisis that has cast a cloud over South Korea's recovery. When that cloud clears, Daewoo (Great Universe) – the name Mr. Kim chose in 1967 when the firm was just a tiny speck in what was then one of the poorest countries in Asia – will remain only on its cars and trading operation, according to the plan. The rest of its dozens of Korean businesses are to be sold off by the end of the year. Daewoo has already announced that an investment group of American banks has agreed to buy most of its electronics business for $3.2 billion. In exchange for all this, the government will prevent the creditor banks from forcing the firm into bankruptcy and liquidation. Despite Daewoo's long-standing financial troubles, led by its $50 billion in debt, the near-collapse of the group in July (1998) still shocked many Koreans. They had believed that Hyundai, Daewoo, Samsung and LG, the four biggest chaebol or family-controlled business groups, were too big to fail." (Yonhap News, 1999)

Its car company, Daewoo Motor, was acquired by the American firm, General Motors. Yonhap News reported: "Finding buyers for Daewoo's businesses will be hard too. The group announced last April that it wanted to sell its shipbuilding operation but has yet to find a taker. Creditors hope General Motors (GM) will take over a controlling stake of Daewoo Motor, but the American car giant is refusing to be rushed into a deal. After a joint venture fell apart in 1992 over Daewoo's plans to enter the American car market, relations between the two companies have been prickly." (Yonhap News, 1999)

To understand the background, let us briefly review the history of Daewoo. The founder, Kim Woo-jung was the son of a provincial governor in Daegu. He was born in 1936, when Korea was still under Japanese occupation. He went to Kyonggi High School and then to Yonsei University in Seoul, graduating with an economics degree. He went to work for a trading firm, which specialized in textiles and clothing. He left and began his own trading firm, which he named Daewoo Industries.

In 1965, President Park Chung-hee had proposed five year plans for Korea to develop export industries, Daewoo followed this strategy with the aid of government licensees to trade and loans from government–controlled banks. Kim Woo-jung bought different companies and turned them into profitable enterprises, growing the Daewoo group. The group included machine tools, construction equipment, ship building, petrochemicals, defense products, civilian helicopters and airplanes, electronics and semiconductor chips, telecommunications products, and musical instruments.

Daewoo Motors produced automobile and became the sixth largest car maker. By 1990, Daewoo was second in all chaebols for value of its assets and third in terms of revenues. But like all the chaebols, growth had been financed by loans. Daewoo's financial base was too highly leveraged, with debt five times the value of its assets.

Yet even in 1997 (the time of the financial crisis) the strategy of Daewoo continued to add even more debt. Earlier debt had been a strategy for rapid growth, but in the time of crisis, debt was a flaw. Yonhap News wrote: "Daewoo's financial trouble came to a head only last month, but its failure was a long time coming. Last year, when the economic crisis forced most of the chaebol to cut back, Daewoo brazenly added 14 new firms to its existing 275 subsidiaries—and this in a year where the group lost a combined 550 billion won ($458m) on sales of 62 trillion won ($51 billion). At the end of 1997, South Korea's four biggest chaebol averaged debt of nearly five times their equity. But while Samsung and LG cut back during the subsequent year of economic crisis, Daewoo acted as if nothing had changed: it added 40% more debt." (Yonhap News, 1999)

In the midst of the 1998 financial crisis, Daewoo added fourteen new firms to its 275 firms, losing a total of $258 million dollars on sales of $51 billion. The following year in 1999, Daewoo went into bankruptcy under debts of $84 billion dollars. Chairman Kim Woo Chung went to France.

When Kim returned five years later, he was arrested and tried. Then Yonhap News reported: "Daewoo Group founder Kim Woo-chong was sentenced to 10 years in jail on Tuesday by a local court that convicted him for massive fraud, foreign currency law violations and other crimes, court officials said. The Seoul Central District Court also ordered Kim to forfeit 21.4 trillion won (US$22.62 billion) and pay a 10 million won fine during the first sentencing trial on the case. The former Daewoo chairman, however, was acquitted of bribery charges, the officials said. 'The heavy jail sentence was inevitable because he ignored corporate ethics and committed irregularities only to lead Daewoo Group to bankruptcy,' Judge Whang Hyun-zoo said. 'This incurred losses to lenders and in taxpayers' money as it forced the government to spend huge public funds to cope with the Daewoo crisis.' (Yonhap News, 2005)

Also Yonhap News reported: "Kim, 69, had been in prison since June last year (2004) after returning home from nearly six years on the run overseas. He was charged with masterminding accounting fraud worth 20 trillion won, illegally borrowing 9.8 trillion won and smuggling $3.2 billion out of the country through the conglomerate 's overseas financial organization, the BFC (British Finance Center). His court procedures were delayed, however, as he underwent heart surgery at a South Korean hospital. He was released in August as the court accepted his request that his detention be suspended for health reasons. The court did not cancel its decision to suspend his imprisonment for one year considering his old age and heart and intestine illnesses. This effectively keeps Kim out of prison until July 28. . . The court said it handed down a lighter sentence than that sought by the prosecution in light of Kim's leading role in assisting the rapid growth of the modern South Korean economy and his business group. 'His contribution to the national economy with extraordinary diligence and passion deserves a corresponding evaluation,' Judge Whang said. . . . Most of the eight other former Daewoo executives who stood trial before Kim were convicted for their involvement in the founder's financial irregularities." (Yonhap News, 2005) Two years later in December 2007, the Korean President Roh Moo Hyun pardoned Kim.

The chronology of the Daewoo failure was this. Daewoo lost 12 billion dollars (12 trillion won) in 1998. In July 1999, twelve of the Daewoo companies went bankrupt. A 'workout' (plan to turn profitable) was agreed with by the Government and Dawoo, with a restructuring of debt of 59 billion dollars into convertible bonds or equity; and on November 2000, Daewoo Motors went into court receivership. (Mako, 2002) The result was that Daewoo had been reorganized into three corporations: Daewoo Electronics, Daewoo Engineering & Construction and Daewoo International Corporation. Daewoo Motors disappeared, acquired by General Motors, which renamed it GM Korea.

INDUSTRIALIZATION – FINANCE AND TECHNOLOGY

Thus the world is changed, even as the world is unchanged. While the contexts of history change, still the underlying patterns of human society continue. Why do some global corporations go bankrupt? How are global corporations threatened in global financial crises? And why is the economic health of nations tied to the economic survival of global corporations? What we will see exemplified in Daewoo's crisis is the complicated interaction between (1) global corporations, (2) global markets, (3) global financial systems, (4) technology, and (5) national economic development.

The coupling of financial crises to industrial disaster is usually direct. For example about the later U.S. 2007 crisis, Nouriel Roubini and Stephen Mihm wrote: "Normally banks issue 'letters of credit' to guarantee that goods in transit . . . will be paid for when they reach their final destination. Once the credit markets seized up after Lehman Brothers failure (a large U.S. investment bank bankrupt in 2007), banks stopped providing this essential financing (for international trade). Global trade came close to a standstill; . . . As one expert on global shipping observed shortly after Lehman's collapse, 'There's all kinds of stuff stacked up on docks right now that can't be shipped because people can't get letters of credit.'" (Roubini and Mihm, 2010)

Technology accelerates financial collapse because of the speed by which capital can now electronically flow around the world. Technology has been and continues to be basic to modern industrialization.

Historically speaking, to date, prosperity in any nation has been achieved through an 'industrialization' based upon technological innovation. But there is a problem here about technology and stability. In a society, innovation is a temporary state; while societal stability needs to be a continuous state. Technological innovation is the great opportunity for societal development, but it also creates a dilemma. Society needs both stability and progress. Yet one is static and the other transient. Can a technological society be stable, achieving a prosperous stasis? This is a profound puzzle. It occurs because of a clash between the finiteness of technology and the infinity of political demand.

So far in history, the realities of industrializing societies have been in either 'prosperity' or 'instability', and alternating. Even with continuing development of science and technology, the 'stability' of a technological society is a current problem (e.g. the Global Financial Crisis of 2007-08 or the Euro Crisis of 2010-13). The challenge for any technological society is between truth and ideology. Technological progress is ultimately conceptually finite. Available natural resources are ultimately finite. This is why generalized but empirically-grounded theory (truth) is the heart of science. Yet political thinking and ambitions can be unbounded, infinite. This is why 'compromise' is the heart of politics, bounding political ambition and political principles. And this is the basic clash between science and politics – between truth and ideology.

These are some of the historical facts at the beginnings of the twenty-first century. Technological innovations continue at a rapid pace. Industrializations of many nations continue at a rapid pace. All this is occurring in what we can now call a 'global civilization'. But nothing therein is stable. Neither prosperity nor security nor justice is stable in the industrializing nations of the world. How can a global civilization develop toward a prosperous, secure, and just world, while preserving the natural environment in which the human species has evolved?

General Motors Bankruptcy

After Daewoo Motor's bankruptcy in 1998, it was acquired by General Motors. But a decade later in the United States, GM itself went bankrupt in a U.S. financial crisis. Bill Vlasic wrote: "Just two months after celebrating its 100th birthday (in 2008), General Motors is facing the grim prognosis that it may not survive to see another year unless it is rescued by a bailout from the federal government. Shares in GM sank to their lowest point in 65 years, to $2.92, on Tuesday, the day after the company revealed in a federal filing that its 'ability to continue as a going concern' is in substantial doubt because it may run out of money by the end of the year. GM said it might be unable to pay its suppliers, meet its loan covenants or cover health care obligations in its labor contracts. The extent of GM's financial crisis, revealed in greater detail in its filing than it acknowledged before, is proving to be far worse than investors and analysts expected just last week. Only an emergency federal bailout seemingly stands between GM and a bankruptcy filing, according to industry analysts." (Vlasic, 2008)

Antony Currie, Rob Cox and Martin Hutchinson wrote: "What drove General Motors into bankruptcy? It's tempting to blame the broader financial crisis, which torpedoed the United States economy, dragged down car sales and shut carmakers out of the capital markets. That overriding factor certainly precipitated the demise of America's largest carmaker, but it was not the ultimate cause. Insolvency has, in fact, been looming over G.M. for several years. In April 2005, for example, while markets were worrying about rating agencies downgrading the carmaker's debt, GM's onerous unfunded health care and pension liabilities looked to have punched a multibillion-dollar hole in its balance sheet. Even then, some suggested that a filing for Chapter 11 protection might be the only way to get GM back on track." (Currie, Cox, Hutchinson, 2009)

For the last twenty years, GM had been immensely profitable, paying its executives large bonuses each year', while not fulfilling its legal (and moral) obligations to its workers, in fully funding their health and pension benefits (as the executives had agreed to do in their labor contracts). Nor had the U.S. government enforced regulation to ensure that large U.S. corporations fully funded their pension obligations. Why? When profitable, GM did not meet its labor obligations; and when unprofitable, GM blamed labor for its bankruptcy. Yet historically, GM executives had not showed any political will to encourage the U.S. government to 'fix' the pension or health care systems – with adequate and properly funded national plans. GM had encouraged a U.S. system of private company pension plans and private company health insurance plans – even while GM executives saw such plans as unworkable for GM over the long term. The result of a long line of GM's executives, lacking financial-integrity, public-good will, or political fortitude, was that GM went bankrupt in 2008 (when global financial collapse occurred).

Currie, Cox and Hutchinson wrote: "That's not to say G.M.'s executives sat back and did nothing. In the last four years they have halved the union work force in the United States; laid off white-collar employees in droves; renegotiated labor contracts and health care expenses;

and sold businesses like Allison Transmission and GMAC, the financial services arm, to raise cash." (Currie, Cox, Hutchinson, 2009)

GM's executives had not fulfilled their obligations to labor when profitable, and when unprofitable, they fired labor – halving their work force and increasing unemployment in the U.S. In Korea, under bankruptcy and unfulfilled public liabilities, Kim and Daewoo executives were tried and convicted for fraud. This didn't happen in the U.S. to GM executives.

Yet just cutting labor while maintaining executive bonuses was not really a solution to GM's basic problem. Currie, Cox, and Hutchinson wrote: "But none of that was enough. G.M. was losing market share faster than it could shrink its operations, even before the precipitous declines of the more recent credit crisis." (Currie, Cox, Hutchinson, 2009)

Roger Lowenstein wrote: "Who shot General Motors? The company's stock is at its lowest level in 50 years. The automaker is weighing yet another round of layoffs — and maybe even a fire sale of venerable brands like Buick and Pontiac. Once General Motors manufactured half the cars on the American road, but now it sells barely 2 in 10. . . . The immediate cause of GM's distress, of course, is the surging price of oil, which has put a chill on the sale of gas-guzzling sport utility vehicles and trucks. The company's failure to invest early enough in hybrids is another culprit. Years of poor car design is another." (Lowenstein, 2008) GM had failed to develop and sell fuel-efficient cars.

As the South Korean government assisted the chaebols in weathering the Asia financial crisis of 1997, so too did the U.S. government assist GM in surviving the Global financial crisis of 2007. David E. Sanger, Jeff Zeleny and Bill Vlasic wrote: "President Obama will push General Motors into bankruptcy protection on Monday, making a risky bet that by temporarily nationalizing the onetime icon of American capitalism, he can save at least a diminished automaker that is competitive. The bankruptcy is a moment of reckoning for an industry that was once at the heart of the American economy. . . . GM's bankruptcy petition culminates a remarkable four months of confrontation between Washington and Detroit that is expected to result in a drastic downsizing of the company. It also places the government in uncharted territory as a business owner, as it takes a 60 percent ownership stake in the company during its restructuring." (Sanger et al, 2009)

But despite government help, GM's executives were not contrite. Micheline Maynard wrote: "General Motors did not apologize for anything in its first trip to Congress more than two weeks ago to plead for a federal rescue. The company's only problem, it insisted, was the current financial crisis. 'What exposes us to failure now is not our product lineup, or our business plan, or our long-term strategy,' Rick Wagoner, GM's chief executive, said in his testimony. On its return visit to a skeptical Congress this week, however, General Motors bowed its head. "GM has made mistakes in the past,' Mr. Wagoner told Congress, and named three: agreeing to expensive union contracts, not investing enough in smaller cars and failing to convert its plants so they could build more than one type of vehicle. It was an unusual concession from a company that has rarely felt the need to apologize for anything, given its bragging rights as the world's largest automaker with operations in 35 countries, and as a company that has built 445 million vehicles and sat atop corporate America for much of its 100-year history. But the mistakes Mr. Wagoner acknowledged do not begin to explain why General Motors finds itself on the brink of insolvency, begging Congress for financial help." (Maynard, 2009)

About the leadership of GM, Maynard wrote: "In recent years, despite many challenges to his leadership of General Motors, Rick Wagoner had managed to keep a firm grip on his job, like hands wrapped tight around a steering wheel. During his tenure as chief executive, beginning in 2000, the company's stock has fallen from $70 a share to less than $4 now, and its market share has fallen roughly 10 percentage points. There have been many challenges to his authority, most notably from the investor Kirk Kerkorian in 2006 and from angry members of Congress during hearings last fall. Throughout the attacks, he had managed to retain the unwavering support of his board. For a time, it seemed he might become the rare chief executive who gets another chance, this time to try to fix many of the problems that occurred on his watch. But he appears to have met his match in President Obama, whose calls for sacrifices from all sides apparently included a call for Mr. Wagoner to step down. In a statement early Monday, Mr. Wagoner said he had been urged to 'step aside' by administration officials, 'and so I have.'" (Maynard, 2009)

Maynard wrote: "Mr. Wagoner joined GM's financial operations in 1977 out of Harvard Business School, and, like generations of executives before him, worked nowhere else during his career. Mr. Wagoner vaulted into Detroit's consciousness in 1992 upon another resignation during a financial crisis — that of Robert C. Stempel, the chief executive at G.M. at the time. Then only 38, Mr. Wagoner became GM's chief financial officer. Two years later, he was named president of its North American operations. His mentor, the chief executive John F. Smith Jr., named Mr. Wagoner president of GM in 1998, and he succeeded Mr. Smith in the top job in 2000." (Maynard, 2009a)

Maynard wrote about Wagner's focus on finance, and not on product, as limiting: "Like Mr. Smith, Mr. Wagoner aggressively expanded GM's operations outside the United States. The company now sells 65 percent of its vehicles overseas, thanks to Mr. Wagoner's push into markets like China, Russia and Latin America. However, GM's sales slump at home led to it losing its longtime title last year as the world's largest auto company, replaced by Toyota. 'It's a pretty unceremonious ending,' said John Casesa, an industry analyst and managing partner of the Casesa Shapiro Group. 'G.M. lost its way in the '70s, but the company didn't know it until 20 years late. The hole was much deeper than he (Wagoner) realized when he became C.E.O. And, Mr. Casesa said that Mr. Wagoner's finance background might have been a poor fit: 'The most successful auto companies are run by people who came out of the revenue-generating functions — manufacturing, design, marketing — making cars and selling cars. Mr. Wagoner . . . skipped the whole apprenticeship that most auto C.E.O.'s experience.'" (Maynard, 2009a)

But why did not Mr. Wagoner's financial skills enable him to foresee GM's financial future? Maynard wrote: "Mr. Wagoner presided over some of the biggest losses in G.M. history. In 2002, the company had predicted that it would earn $10 a share by the middle of the decade. Instead, G.M. lost $30.9 billion in 2008, when its per-share loss translated to more than $50 a share. G.M. stock, an economic bellwether that sold above $35 only three years ago, closed Friday at $3.62; it has fallen as low as $1.27 in the last year. In 1994, when he took charge of GM's North American operations, the company made up 33.2% of auto sales in the United States. Last month, G.M. represented only 18.8 % of American car and truck sales, according to statistics from Motor Intelligence, which tracks industry data." (Maynard, 2009)

A decline in market share should be a clear indication of a problem in a company's product lines. Six years after taking charge of GM North American operations, Wagoner did finally realize GM's products were a problem. Maynard wrote: 'Under pressure to stop GM's sliding market share, Mr. Wagoner hired Robert A. Lutz, a longtime auto industry executive, in 2002. Mr. Lutz reorganized GM's product development operations, and introduced a number of new vehicles, including sporty models like the Pontiac Solstice and Saturn Sky. Both Mr. Lutz, who had previously announced his plans to retire by year's end, and Mr. Wagoner have championed the Chevrolet Volt, a plug-in hybrid electric car that G.M. plans to introduce in late 2010. Mr. Wagoner has said that one of the moves he regretted most was GMs decision to kill the EV-1, an electric car that it leased to customers in the late 1990s. Although the vehicle was not profitable, it helped GM's image with environmentalists, which in 2006 Mr. Wagoner conceded he had understood too late." (Maynard, 2009)

Foresight and strategy are two major requirements for good leadership; but neither appeared to have been GM's executive strength. Maynard wrote: "Only six months ago, Mr. Wagoner stood in front of hundreds of G.M. employees in the atrium of the company's Detroit headquarters, celebrating the automaker's 100th anniversary. Dressed in a gray suit and a yellow, blue and white striped tie, Mr. Wagoner said: "So, what's our assignment for today and tomorrow? Above all, it's to demonstrate to the world that we are more than a 100-year-old company. We're a company that's ready to lead for 100 years to come." (Maynard, 2009) But GM was not ready and in six months went bankrupt.

GLOBAL COMPANIES

Comparing the vicissitudes of Daewoo and GM, one can see several common features about global companies (some of which are good and some not so good).

1. **Global companies are focused upon exporting from their home nation.**
   Exports are important to any nation, providing foreign currency to purchase needed imports for a nation.
2. **Global companies value foreign markets as much as their home market.**
   Even in the financial crisis, Daewoo looked to increase exports from Korea into the American market. Wagoner, as CEO of GM, focused upon sales of GM cars abroad, even as GM's home market share declined from 33% to 18%. Wagoner had 65% of GM production exported.
3. **Both global and national market shares depend upon the quality of the products of a global company.**
   GM's market share decline was due to the poor quality of their product design; and its bankruptcy occurred when their product line was constituted of fuel-inefficient large vehicles and then gas prices dramatically rose. Korean cars were fuel efficient and developed, over time, a high-standard of quality.
4. **Global companies are dependent upon the financial stability of their home nation.**
   Both Daewoo and GM became bankrupt during a financial crisis which plunged their home nations into depression.

5. **Global companies are dependent upon home nations for educational and social infrastructures; and nations are dependent upon Global companies for technological innovation.**
   Both the U.S. and Korea provided schools and universities to educate labor and executive personnel for industry. The U.S. government did not provide sufficient national-based pensions nor a national health-care system, transferring this social service to industry (which, in turn, failed to fully fund their labor-contract obligations – resulting over time to burdensome social liabilities to the company). The U.S. government support of science at U.S. universities provides scientific depth for Global companies with research laboratories in the United States. Korean government support of scientific research in laboratories and universities provide science bases for Korean chaebols. Both in the U.S. and in Korea, new technology is implemented in the innovation of new high-tech products and services by Global companies. Global companies, worldwide, are dependent upon home nations for scientific capability; and all Nations are dependent upon global companies for technology innovation.
6. **Nations are dependent upon global companies both for their contribution to GNP and to national employment.**
   Both Korea and the U.S. nations were dependent upon home global companies to provide significant contributions to their respective GNPs and levels of employment.
7. **Global companies care more about their executives than their laborers.**
   GM provided large executive bonuses when profitable but did not fully fund obligations to GM laborers; and when profits declined, GM cut fired half its labor force but not its executives. In Korea, after severe industrial strikes, chaebol management came to terms with industrial unions.
8. **When a global company faces bankruptcy, it appeals to its home nation for a bailout.**
   Both Daewoo and GM turned toward government assistance to bail them out of cash-flow crisis due to excessive liabilities, when sales collapsed in a year of crisis. Daewoo was restructured by Korean government help; and GM was downsized by U.S. government help.

GLOBAL CIVILIZATION

The goal of historical studies is not only to understand the past but also to draw lessons for the future. But how can one empirically study for the future? One can project trends and patterns from the past, knowing that each is context dependent and will not occur exactly as before. The methodology is to identity the underlying reasons which drive the patterns of societal activties. Underlying any pattern are societal factors which will continue to drive society in the future, just as they have in the past – driving forces and factors that underlie societal forms. This is the essence of historical analysis, to discover and clarify the underlying drivers of human history. And when society changes dramatically, this kind of envisioning possible futures is important. Otherwise, society is blindly driven into the future by forces (such as technology, governance, resources, economic systems, etc.), which narrow options and cause grief and hardship.

We have been using the term of a 'global civilization'. What are the features of this concept, and why is it an apt description of the future of the world for the twenty-first century? There was a particular time which divided our international past from our global future. This occurred in the year 1995, when a powerful new technology began driving societal patterns all over the world. It was the Internet, and 1995 was the year of its commercial innovation. This innovation had immediate impact upon economies, first that of the U.S. economy and then of the world. The impact on the U.S. economy of the Internet (a critical tool for globalization) was immediately seen in a U.S. financial boom from 1995 to 2000 – the now famous, or infamous, "dot.com" financial bubble of the U.S. stock market, Figure 1.1.

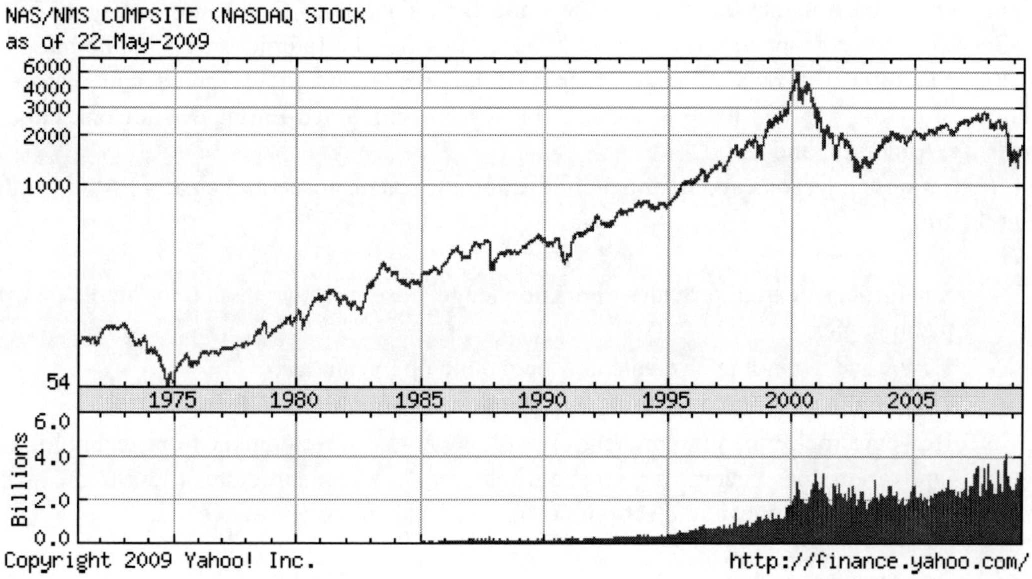

Figure 1.1. U.S. Stock Market "Dot.Com" Financial Bubble.

The Internet had been invented as a technology to enable computer networks, created from scientific research in computer science. It began as ARPAnet in 1965 (Advanced Research Projects Agency network, funded by the U.S. Department of Defense). In 1985 the network was transferred to NSFnet (National Science Foundation network, funded by the U.S. National Science Foundation). In 1995, this was transferred to the commercial sector as the 'Internet'. The Internet technology had gone from universities and government to industry.

This last transfer was coincident with the U.S. government deregulating the U.S. telecommunications business. It was in both that deregulation and the new Internet that many entrepreneurs saw new business opportunities. They launched both new telecommunication businesses and many new businesses in electronic commerce, "dot.com businesses". After 1995, the rapid growth of the Internet in the U.S. was very exciting. For example in 1996 in the United States, 14% of the population used the Internet, which jumped to 22% in 1997,

31% in 1998, 38% in 1999, and 44% in 2000. In 2000, the average monthly hours-spent-on-line-per-user was 19 hours. U.S. consumer spending online grew from a few million in 1996 to $3 billion dollars in 1997, $7 billion in 1998, $19 billion in 1999, and $36 billion in 2000. Of the $36 billion spent online in 2000, $11.0 billion was for travel, $7.7 billion for PCs, $2.4 billion for clothes, $13.4 billion for books, and $13.4 billion for other merchandise. The Internet stimulated an economic expansion in the United States economy.

Venture capital funds poured huge amounts of capital into hundreds of ideas for new dot.com businesses, and immediately took many of these public as IPOs (initial public offerings). From 1997 to 2000, the excitement over the Internet and the new dot.com businesses drove U.S. stock markets to new highs. Business pundits then wrote that the Internet was creating a "new economy", for which one need not need worry about profitability of a new business but market growth. But in the year 2000 this stock bubble burst. The so-called "new economy" fell back to the principles of the old economy, which require profitable businesses. Many new unprofitable dot.com businesses collapsed. The NASDAQ average dropped 65% early in 2000. Following the Internet stock bubble bursting in 2000, there was a rash of major corporate failures – due to the greed and misdeeds stimulated by the bubble. Some of the failed companies included Enron, WorldCom, Global Crossing, Andersen, and so on.

The important lessons about global civilization that one can see in the early history of the Internet are:

- Scientific research provides the knowledge base for the invention of new basic technologies.
- New basic technologies create new economic opportunities.
- The path to wealth in economic opportunities is often financially rocky.
- However rocky the entrepreneurial path, economic development from technological innovation does become a permanent feature of societal structures (despite the stock bubble, the Internet and electronic commerce was in the world to stay).

Science and technology and innovation continue in the world's industrialization as the heart of 'global civilization'. From 1995, the Internet spread globally and became the hallmark of modern economic system, some have called 'knowledge economies'. Of course, there has always been an 'international world', since the social forms of empires and nations emerged (even back in ancient times). But in the twenty-first century, the term 'international' was too weak to capture the 'economic intensity' of competition in a new global civilization. And the concept of 'global civilization' centers fundamentally not only upon national survival and but also survival of the entire planet.

There are six factors in the societal pattern of globalization.

1. A global civilization is one in which commodity and financial markets in the world are interconnected.
2. It is a civilization in which global corporations dominate economic production and trade in the world.
3. It is a civilization in which culture in shared through global information media.

4. It is a civilization of finite natural resources upon which infinite global demand is made.
5. It is a civilization struggling with the meaning of two core societal concepts – democracy and capitalism.
6. It is a civilization of limited insurrection, overshadowed by nuclear Armageddon.

This is the question. How do nations survive and prosper in this global environment?

NATIONS AND GLOBAL COMPANIES

We see in this historical comparison, something about the differences in the two major players in modern global civilization – nations and global companies. We also see their interdependence for the economic vitality.

- For Nations, their primary interests lies in exports and imports, employment, resources, social justice, governance, taxes, societal infrastructure.
- For Global Companies, their primary interests lie in markets, technology, capital, profits, executive careers, stock markets, finance.
- Nations are dependent upon Global Companies for their contributions to national GNP, employment, technology, products, exports, taxes.
- Global Companies are dependent upon Nations for social infrastructure in education, research, health, security, pensions, market regulations, financial systems, legal systems.

Thus the two global players (global companies and nations) are complementary to one another. With proper cooperation and partnerships, national economies and the global economy can prosper. But without such, national economies can collapse and global companies can go bankrupt.

What we wish to learn from the Korean experience, concretely and empirically, is – what kinds of cooperation and partnerships between industry and government worked and what didn't work. In the Korean experience of the new global civilization, what were the opportunities and what were the problems?

Also central to the cooperation between government and industry are 'individuals' in positions of power and their 'characters' in leadership. As we will see in the case of Korea, exceptional leadership was shown by government and industrial leaders!

For example, Kim Woo-chong's leadership was outstanding (until his strategy of too-much financial leverage tripped Daewoo in the Asian financial crisis). One can gain some insight into Kim's character from his autobiographical book. He wrote about the time when he was young during the Korean war: "With my father kidnapped and taken north and my older brothers in the army, I was, at age fourteen, responsible for feeding the family. I had to sell at least a hundred newspapers in the market a day to cover basic expenses; my mother and younger brothers waited late into the night for me to come home so we could all eat together. I was always very grateful to them for that. I really felt great when the four of us were together. I was so happy, and we really enjoyed those meals. But we could not always eat as a family. The Pangchon market in Taegu where I sold the papers was largely makeshift shacks along the river, and often the shops would close during inclement weather. I had to sell my one hundred papers if we were to eat, so bad weather meant trouble. On some days I had only a few extra coins to give the family for the day's meals on my way out in the morning. When I came home on those nights, my mother and brothers would be sleeping, and I quickly came to realize why they were sleeping: there was only one bowl of rice, and they were saving it for me. My mother would get up to feed me and say, 'We've already eaten. You must be hungry, so hurry up and eat.' I thought I would cry when I looked at my brothers, whom my mother had to put to bed hungry because there was only one bowl of rice. . . . We were materially poor, but we were rich at heart. We had very little, but gave to one another what we could. Someone who has everything in the world but does not know how to give is not rich." (Kim, 1992)

This was a frequent and widely shared quality of the Korean people, 'materially poor but rich at heart'. After being occupied as nation and then put though national partition and civil war, they struggled toward national prosperity and security. This is a first lesson we will learn, about the global world, from the success of the South Korean people. They were materially poor but rich in heart – and rich in leadership.

Looking Ahead

The general story of Korea's successful economic growth is amazing, and we will not recount it in detail. Instead, we examine the critical elements in the story – about why the success was possible. Briefly, the story is, as Anne O. Krueger and Jugho Yoo wrote: "Korea's spectacular economic growth from the 1960s to the 1990s is well known. Korea was one of the poorest countries in the world, widely regarded as a country without serious growth prospects. After economic policy reforms began in the early 1960s, Korea began growing at sustained rates previously unheard of in world history. Real GDP (Gross Domestic Product) grew at an average annual rate of 10 percent per annum in the decade starting in 1963." (Krueger and Yoo, 2002)

This is a critical element. Why did 'economic policy reforms' help jump start Korea from 'one of the poorest countries' to 'previously unheard' of growth? What were these policies and how implemented?

Both the government and private industrial firms were responsible for their successful implementation. Krueger and Yoo wrote: "First steps in reform included moving to a more realistic (and constant real) exchange rate for exports, and the relaxation of restrictions on importing by exporters. . . . Other major reforms also took place, including a major fiscal and tax reform in 1964, gradual removal of price controls, and a shift from a regime discriminating against agriculture to a protective one." (Krueger and Yoo, 2002) The South Korean government policy toward encouraging exporting industries was important, contributing to South Korean economic growth. Also the South Korean government policy toward protecting agriculture was also important, as this would contribute to South Korean social justice. In 1963, 90% of the population still existed on farming.

Central to private business was the South Korean form of the 'chaebol'. Krueger and Yoo wrote: "The chaebols were successful exporters and for the first decade or more of Korean growth were regarded almost as the 'heroes' of Korean development. . . . The chaebol were generally encouraged to enter whatever export markets they could, but when the authorities wanted a venture undertaken, the chaebol were asked to do so. . . They undertook these ventures with the implicit guarantee of the government that credit, tax exemptions, and other support would be available to make the venture profitable. But the chaebol were on the whole remarkably profitable and had little difficulty with servicing their debt." (Krueger and Yoo, 2002)

Thus some of the critical things we need to explore include:

- The nature of the partnership between South Korean government and South Korean private industry.
- The nature of South Korean government policies, which encouraged exports (but did not create 'state industries').
- The nature of leadership in South Korean government and in South Korean chaebols – in terms of competence, integrity, and patriotism.
- The nature of South Korean fiscal policies and of International financial systems.

As we have noted, the Korean story has been neither simple nor without difficulty. Kureger and Yoo wrote: "Thus by the mid-1990s, Korea had sustained three and a half decades of rapid growth. While there had been periods of difficulty . . . Korean policy makers had met their challenges successfully. . . . (Also) the late 1980s had witnessed the introduction of democracy into Korea. The elected governments chose to liberalize further, including especially the financial sector and international capital flows. . . . (But) underlying weaknesses were not addressed, and the stimulus to the economy, through the expansion of domestic credit . . increased the vulnerability of the financial system . . . . A number of events took place early in 1997 that surely eroded confidence. One of the large chaebols, Hanbo, went bankrupt early in that year. . . (International) capital flight began early in the fourth quarter of 1997. . . (In November of 1997) by the time the International Monetary Fund (IMF) was approached (Korean Central Bank) reserves were being depleted . . . approaching zero within 48 hours." (Krueger and Yoo, 2002)

The South Korean story is not merely a story of a developing nation but also a dramatic story of competition in the new global civilization. It was the transition of the world from an international economic system to a global system which had created opportunities for Korea's miracle growth but also challenges to its continuing stability.

It is into this well-known Korean story that we will delve – but deeply, to look at the critical factors in success. To do this, we will first examine how three of the major Korean groups became dominant global companies: Samsung, Posco, Hyundai. We next look at the history of government industrial policy in modern Korea. And we look in detail at the impact of the global financial system on the Korean economy. Then we turn from looking at the global companies (chaebols) in Korea to small and medium sized (SMEs) businesses. We look at Nano, a new high-tech small business in Korea growing to become a medium-sized business with international competitiveness. Finally, we summarize what can be learned from the Korea experience – about proper industrial and fiscal policies for national prosperity in a global economy – and about the pitfalls to the nations of global civilization.

And we will look at this case from more than a mere economic perspective. We will use a broader cross-disciplinary perspective, which uses all of the social sciences concepts (and not just economic concepts). Because of this broad perspective, we will find empirical evidence for some important and general lessons about economic development.

What we will see in our analysis of the development of South Korea is empirical evidence for the following postulates:

1. Communist/socialist ideologies were wrong, creating only inefficient and starvation economies, under unjust and brutal dictatorships.
2. The two fundamental challenges about industrialization in the world are economic competitiveness and democratic justice. Economics and politics working together are necessary for a prosperous and secure society.
3. 'Laissez faire' economic ideology (which emphasized economics but demeaned politics) has been incomplete for effective economic development and democratic growth in an industrializing society. A free market economy is necessary for economic growth, but there is a proper role of government in economic development. Government needs to construct a proper context for economic activity – including an effective societal infrastructure, effective regulation of free markets (to insure safe products), a stable financial system, and just rule.
4. Accordingly, traditional macro-economic theory has proven insufficient for policy guidance of national economic development. A complementary contextually-dependent micro-economic model is needed to supplement macro-economic theory, modeling the real societal contexts of nations.
5. Corrupt leadership distorts and limits economic growth; and unjust societal structures impede and disrupt economic growth. Both good leadership and proper societal structure are necessary for effective economic growth. But how are the conditions for the 'good' in leadership institutionalized? And how is corruption in a social structure limited, identified and corrected?
6. In a global world, national economies are regionalized; but national financial systems are internationalized. How can international regulation of a global financial system be created?
7. Even in a global civilization, economic limits still are bounded by natural resources. What technologies and political structures are necessary to create sustainable national economies, which do not destroy the individual and society, nor the planet?

Chapter 2

Discontinuities in Global Industries – Samsung

Introduction

In the success of the development of South Korean national competitiveness, an important factor to examine is technology. How from a historically impoverished and occupied background, did the new nation so quickly rise to technology competitiveness? Samsung Electronics illustrates that achievement, only entering electronic semiconductor production in 1981 and yet, by 2010, dominating the world memory chip market.

Innovation has driven economic development and business competitiveness; and in the past centuries, innovation has occurred principally in national and international contexts. But in the twenty-first century, innovation is occurring in a global context. While many innovation theories abound, the real test of validity is an empirical success in using innovation to succeed in global markets. From 1950 to 1980, the United States was the principle success story for innovation and competitiveness – innovating and competing in computers and electronics (particularly transistors and semiconducting integrated circuits). But from 1980 to 1989, Japan became the principle success story for competiveness, but not necessarily innovation. The U.S. still led in innovation in computers, in the Internet, and in biotechnology. But the U.S. no longer led in long-term commercial competitiveness, being principally focused upon short-term profits. In the 1980s, Japan led the world in commercial competitiveness; and in the 2000s, Korea led in competitiveness. Why? And how?

The world changed. In 1980, China began economic reform within a communist party framework. In 1989, the Soviet Union collapsed from dictatorship communism. In 1985, South Korea began competing global markets. In 1990, Japan stumbled financially. In 1990, the U.S. outsourced its much of its manufacturing and turned its economy toward principally financial services. Yet in the U.S., its financial services (deprived of traditional investments in manufacturing) resorted to a gigantic and fraudulent scheme in mortgage securitization and collapsed the world's banking system in 2007.

Thus by 2010, Korea became the exemplar of success in global competiveness. What strategies had successful Korean firms implemented to achieve global stature? In 2012, Samsung became the largest electronics company in the world, dominating markets in memory semiconductors and displays. Hyundai became the fifth largest automobile manufacturer in world. POSTCO became the fourth largest steel producer. We will examine these cases of successful Korean companies to understand modern innovation theory, in the context of globalization. This is a world economically consisting of global corporations, but still with national innovation systems and also finite global resources.

HISTORY: KOREAN INNOVATION PATH

The case of Samsung occurs within modern South Korean history. In 1963, a new government in Korea under General Park focused upon basic economic development beginning with chemicals and steel. In 1970, government focus turned to high technology and encouraged chaebols to enter consumer electronics. In the middle of the 1970s, many Korean firms were interested in producing IC chips; but this was commercially risky because of large capital investments. Samsung made investments in producing chips, with its production of a 64K DRAM chip in November 1983. But like other Korean firms in the 1980s, there was still had a long way to go capture world markets. IN 2010, Samsung did achieved domination of the world memory chip market, with over 50% share.

We recall that from 1963 to 2012, Korea had achieved remarkable economic performance. Its Gross Domestic Product (GDP) per capita rose from a very low $79 (U.S. dollars) in 1960 to $20,045 per person in 2007. Exports leapt from $33 million in 1960 to $371 billion. What strategy had South Korean firms followed to achieve such economic growth and strong competitiveness in so short a time? Korean firms had an innovation strategy as a path sequence: first acquiring technology, then competing in technology and then leaping-ahead in innovation: "path-following," "path-revealing," and "path-creating." (Choi, 2010) These occurred in successive decades with 'path-following' from 1960 to 1980, 'path-revealing' from 1980 to 2000, and 'path-creating' beginning in 2000.

'Path-following' began in the 1960s, when Korea's economic objective then was to lay the foundation for industrialization – through the development of industries that would transform the import-oriented market structure to the export-oriented light industries (such as textiles and clothing). This expansion of labor-intensive industries created a demand for machinery, raw materials, and components, which were mainly imported. In the 1970s, the Korean government adopted a series of promotion policies to facilitate the growth of local machineries, chemicals, metal products, and basic metal industries. This resulted in generating a huge demand for technologies, which were not available from domestic sources. Then Korea's science, technology and innovation (STI) policies started promoting the inward transfer of technologies from foreign sources and developing the domestic capacity to assimilate and improve the transferred technologies. The Korean government took a restrictive stance toward direct foreign investment, and relied upon long-term foreign loans to finance the selected industrial investment. This led to massive importation of foreign capital, goods, and turnkey plants. Industries acquired appropriate technologies and increased technological capabilities through reverse engineering.

'Path-revealing' began in the 1980s when the Korean government turned to promoting 'high-tech industries' – in order to join the ranks of the advanced countries. Industrial transformation focused upon high-tech areas, such as information technology, biotechnology, and new materials. Korean economic development wanted these more sophisticated technologies, but foreign sources became increasingly reluctant to transfer technologies to Korean industries. The Korean government responded by research and development – developing national R&D programs and promoting private industrial R&D activities through fiscal and financial incentives. In 1982, a national R&D program for promoting target areas was launched by the Ministry of Science and Technology. Following this national program, a series of other national R&D programs were promoted by other ministries. Also the R&D investment of the private sector rapidly increased. Some private enterprises later emerged as international major forces in R&D as technology innovators.

'Path-creating' began in the mid-1990s when the private Korean private firms started playing the role of the major national economic drivers (succeeding the earlier governmental stimulation efforts). Then government policies shifted away from stimulation and toward establishing sound infrastructure and promoting national R&D. By 2012, Samsung dominated the global market in semiconductor memory and display. Hyundai was the fifth largest automobile producer in the world. POSTCO was the fourth largest steel producer.

Korean national economic development policy had encouraged a strong research capability, focused upon a Korean economic development for export-oriented growth. The highly motivated private sector was led by global high-tech chaebols, backed by strong government support for technological capability building.

HISTORY : SAMSUNG BEGINS

The Samsung Group, a Korean chaebol, was founded by Lee Byung-chull. He was born in 1910 into a wealthy Korean family owning land and was of a branch of the Geongju Lee clan. But Korea was under Japanese rule, having been annexed in 1910. From ages 5 to 10, he attended a private school founded by his grandfather and studied Chinese Classics. He then went to the Jisu Middle School in Jinju and later transferred to middle schools in Seoul, where studied modern sciences. He married Park Doo-eul in 1926. In 1930, he left for Japan to study economics at Waseda University. He did not complete a degree but returned to Korea, at the death of his father. He used his inheritance to build a rice mill; but it was not too profitable. He next established a trucking business in 1938 and traded rice and other commodities. He named company, Samsung Trading Co, which meant 'Three Stars'. In 1939, he used profits to take over the Chosun Brewey. (Samsung Foundation, 2012).

Even under Japanese rule, Samsung Trading grew to transport goods throughout Korea and, by 1945, had become one of the ten largest trading companies in Korea. Korea became independent from Japan in 1945, and Lee chartered his trading company as Samsung Corporation. Then in 1950, the North Korean army invaded South Korea, and Lee was forced to move his company from Seoul south to Busan. From there he provided supplies to the U.S. troops entering to drive back the North Korean army. In 1953, at the end of hostilities of the Korean War, Lee expanded by establishing the Cheil Sugar company. With profits from trading and from beginning also from sugar, Lee next founded Cheil Textiles in 1954. Then he used his growing revenues to continue buying and establishing companies – in insurance, banking and other industries. But he was not in any high-tech industry.

By 1960, Lee was one of the wealthiest man in Korea. A student revolution brought down the Syngman Rhee government, which they thought corrupt. It was followed by a second government, still ineffectual, and then replaced by a military coup under General Park Chung-hee. Lee did not get on well with Park. Tony Mitchell wrote: ". . . government finance favored chaebol which promised to build the next generation of Korean industry. Samsung was still at the back of the line at this stage. Huge state-planned projects in shipbuilding, oil refineries and petrochemical plants and machinery sectors went to rivals such as Hyundai, LG, SK Hanwha, and other groups." (Mitchell, 2010)

One area in which Samsung was able to get a government license was high tech electronics. In 1969, Lee established Samsung Electric Industries to produce black and white televisions and electrical appliances. Samsung was able to import technology and parts for TV production from Japanese firms. But TV by then were being constructed with transistors, as Japan led in innovating transistors into consumer electronics. After paying for imported components, there was little profit left in the assembly of the TV.

In 1975, Samsung entered the semiconductor business buying a small firm called Korea Semiconductor, begun a year earlier to produce chips at the LSI scale (16K memory). But it had not sufficient capital to succeed. Lee changed its name to Samsung Semiconductor; and with sufficient financing, it succeeded in producing an LSI scale chip for watches. Youngrak Choi wrote: "But in 1983, Lee wanted to set up more concrete technological bases in Samsung while he was alive. He was already over seventy when it was decided to enter into the semiconductor area. . . . He regarded the semiconductor area as his last business in his life at Samsung. . . He was ready to accept all kinds of difficulties which this firm would face in . . . its large-scale investment in the semiconductor sector. . . Samsung aimed at being a leading firm in Dynamic Random Access Memory (DRAM) chip production in the world as early as possible." (Choi, 1996)

Lee decided to go into semiconductors because he saw clearly (as had all the industry) the exciting technology vision for the future of Integrated Circuit (IC) chips. The challenge was to execute in the face of fierce competition. Also Lee had witnessed the Japanese firms destroy U.S. memory-chip firms at a technology discontinuity in 1981. The race was in successive technology discontinuities, and in the Schumpeterian creative destruction of competitors. To compete and conquer, Samsung would have to catch-up, leap-ahead, and then denominate. Lee's technology strategy was clear (successive generations of technologies) and his business strategy was clear, path-follow, path-reveal, path-create. He was determined and could fund long-term technology investment through the capital structure of a chaebol – cash-flows within the chaebol, one company funding investments in another company.

THEORY: CREATIVE DESTRUCTION IN INDUSTRY

We will continue our case of Samsung electronics. But to appreciate the technological context which Lee faced in 1969-1983, we need some background – three backgrounds, actually, in: (1) the theory of creative industrial destruction, (2) the history of innovation in semiconducting electronics, and (3) Moore's Law as a technology roadmap.

Theory is important to understand as to what lessons can be generalized from particular experiences. To learn from the case of Samsungs' competitive success, we need to review the theory of innovation impacting the dynamics of an industry. To explain a connection between innovation and economic structure, Joseph Schumpeter in 1942 introduced the term 'creative destruction' to describe how a radical technological innovation can disrupt industrial structure, at a discontinuity. (Schumpeter, 1942) Later, David Ford and Chris Ryan in 1981 proposed a chart of market-volume-over-time for an industry to reflect the underlying maturation of the core technologies of its product; and they called it an 'industrial life cycle' (Ford and Ryan, 1981). William Abernathy and K. B. Clark in 1985 described how in the early days of this pattern, the product innovations peaked and then were followed by production innovations. (Abernathy and Clark, 1985) One of the authors in 1988, combined the industrial life cycle with product technology S-curve and numbers to industrial competitors, to show their relationships, Figure 2.1. (Betz,1988).

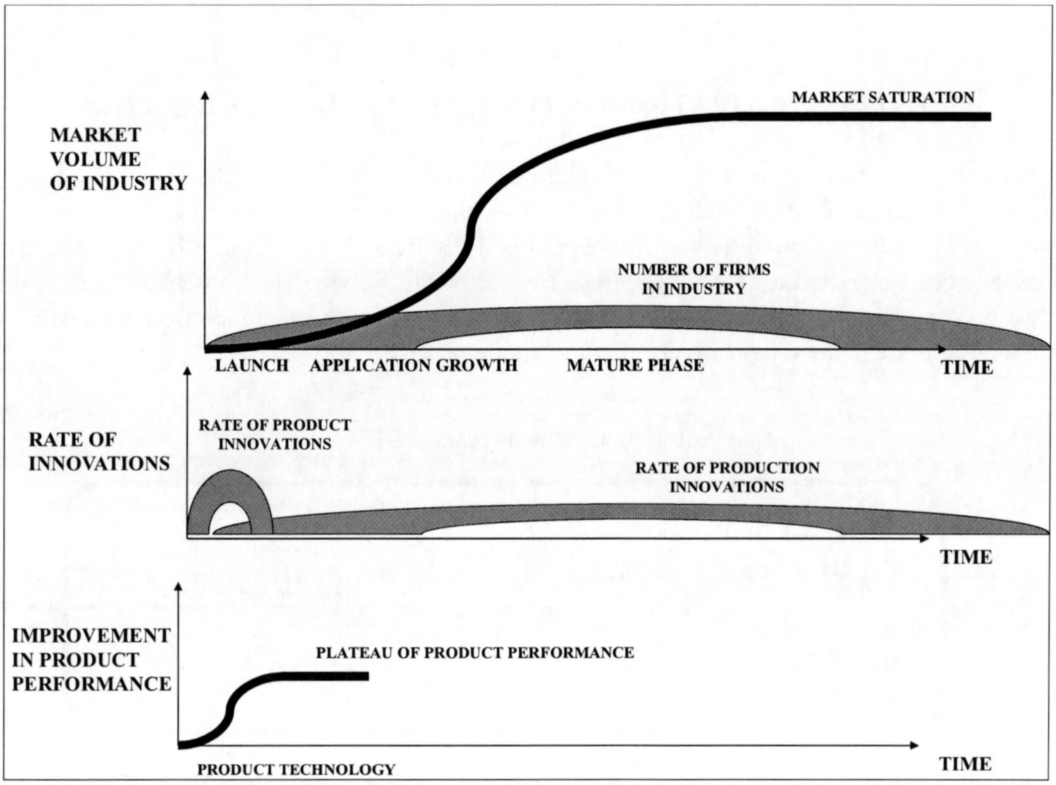

Figure 2.1. Dynamics of New Industries Based on Technological Innovation.

Based upon a new radical technology, new products create a new industry (e.g. the invention of the stored program electronic computer created the computer industry in 1950). In the launch phase, many new product technology innovations spur the growth of the new industry. But after a product design-standard is achieved, the rate of product innovations decline; and innovation shifts to a high rate of production innovations, to produce the products, reducing costs and improving quality. As technical innovation both in products and production slows, then the industry enters a mature industrial phase, with commodity-like products. All products of competitors are similar in performance and cost, with differences principally in product design between competitors. The number of firms in the new industry increases rapidly in the launch phase to peak in the application growth phase and then decline in mature industry phase.

This pattern does capture historical pictures of new high-tech industries, but only when there are no technology discontinuities. Yet the history of the IC semiconductor chip industry is one of several technology discontinuities as 'next generation of technologies' (NGTs), and so this industrial-life-cycle pattern has not been historically accurate for the electronics industry. To revise the life-cycle theory to fit an industry with many technology discontinuities we will review two cases that follow of Schumpeterian creative destructions in the semiconductor IC memory industry – in 1981 by Japanese firms and in 2010 by Samsung. Joseph Schumpeter called the situation of an older industry replaced by a newer industry due to new technology as 'creative destruction' in an economy (technology creating a new industry, while destroying the older technology-obsolete industry).

History: Innovation of the Electron Vacuum Tube

Electronics history is essential background for understanding Samsung's strategy. Electronics began in 1906 with the invention of the electron vacuum tube. Electrons, the particles in electronic circuits, was discovered in 1894 by J. J. Thomson, a British physicist. Lee de Forest invented a device in 1906 – to control the flow of electrons in an circuit, the electron vacuum tube. Electrons were sent through a vacuum in a glass tube from a negatively charged cathode plate to a positively charged anode plate, Figure 2.2.

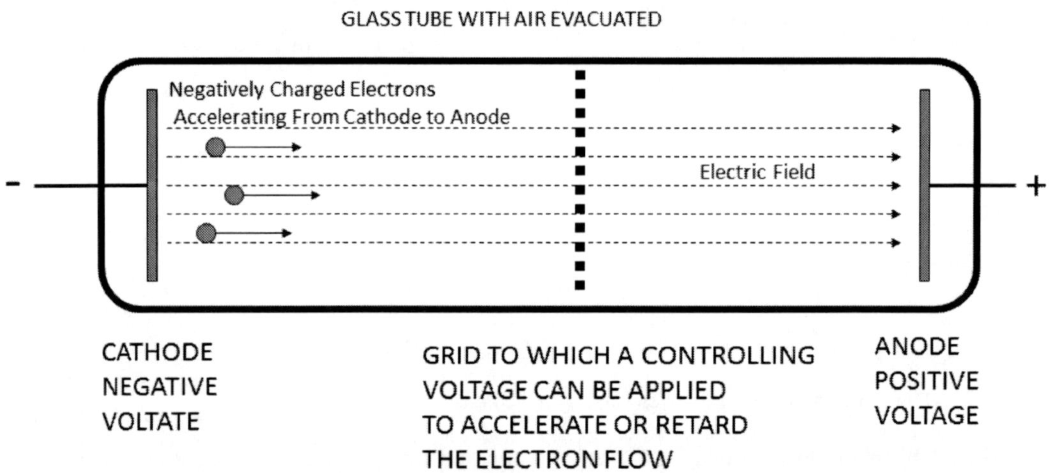

Figure 2.2. Physics of the Path of the Flow of Electrons in a Vacuum Tube.

The intensity of the flow of electrons could be modified by a 'grid' (a mesh of wires between the cathode and anode) to which a controlling voltage could be applied. In this way an electron vacuum tube allowed the control of electronic signals within an electronic circuit. This was the beginning of modern electronics. The electron vacuum tube made possible many subsequent electronic inventions, including radio and television.

But the excellent invention still had technical limitations. The tubes were relatively big and used a lot of electricity to heat the anode (to boil off the electrons). By the middle of the twentieth century, there was a need for an electronic control device better than an electronic vacuum tube. The research laboratory which took upon itself this task to replace the electronic vacuum tube was Bell Laboratories in the U.S.

HISTORY – INVENTION OF THE GERMANIUM TRANSISTOR

Early in the twentieth century, AT&T became the centralized U.S. telephone company to provide long-distance telephone service to the nation. Then engineers and scientists in Bell Labs had leapt upon the invention of the electronic tube and innovated the first mass production of electron vacuum tubes. AT&T used them for amplification of phone conversations. Yet in the 1930s, only twenty years after the innovation of the electronic vacuum tube, some of Bell Lab's scientists were thinking of replacing the electronic vacuum tube by a better device. But what? This was a challenge that Mervin Kelly of Bell Labs undertook in 1936. Then Kelly had been appointed director of research of the Bell Labs and established a research group to search for a way to replace the electron vacuum tube (Wolff, 1983).

Kelly was trained as a physicist, and he had followed the exciting developments in physics of the new quantum mechanics. Kelly saw a technological opportunity in quantum mechanics, as it was providing theory to explain the behavior of electrons in materials. Kelly envisioned that the new theory applied to certain semi-conducting materials, germanium and silicon, might provide a way to invent something to replace the electron vacuum tube. Earlier when radio was being developed, one of the other inventions then was the discovery that germanium could be used to detect radio signals. Also it was discovered that silicon crystals could convert light into electricity and change alternating current to direct current. Both these interesting materials had been called "semi-conducting" materials, because their electronic conductivity was between the high conductivity of metals and the non-conductivity of dialectic materials.

In 1936, Kelly created a research program in Bell labs in solid-state physics, to explore the use of the materials of germanium and silicon with the new theory of quantum mechanics. One of the new hires Kelly made for his new program was William Schockley, who had just received his Ph.D. in sold-state physics from MIT. Michael Wolff wrote: "Kelly gave me (Schockley) an eloquent pep talk. . . . He pointed out that relays in telephone exchanges caused problems . . . . He felt electronics should contribute. . . Dr. Kelly's discussion left me continually alert for possible applications of solid-state effects in telephone switching problems." (Wolff, 1983, pp. 72-73). Kelly's support and Shockley's alertness paid off – but after World War II. During the war, Kelly was responsible for 1200 military research programs performed at Bell Lab. Schockley spent the war at the Pentagon, working on military problems.

After the war, Kelly reestablished the solid-state physics research program and invited Shockley back to Bell. Kelly appointed James Fisk to assistant director of physical research at Bell and told him to make a major and broad research effort in solid-state physics and theory. Fisk set up three research groups: physical electronics, electron dynamics, and solid-state physics. William Shockley and Stanley Morgen led the solid-state physics group. They hired John Bardeen and Walter Brattain, new solid-state physicists, into the group.

Shockley's group learned how to alter the conductivity of germanium by adding other atoms (doping) to the germanium crystal during its growth. Next Shockley proposed an artifact (transistor) to control the current in doped-germanium crystals – by applying an external electric field perpendicular to the direction of current in the crystal. This was to work similarly to how a voltage applied to the grid of an electron vacuum tube controls the flow of electrons in the tube. The reason Shockley thought this was possible for germanium was that he had used quantum theory to calculate the expected effect of the controlling field on the electron flow. His calculations showed that it would be large enough to provide amplification. Amplification meant that the electron flow through germanium would closely follow the increases and decreases of the controlling field, thereby mimicking its form and but with larger amplitude. Shockley made a sample, but it didn't work, didn't amplify. Shockley asked Bardeen, a recent graduate in physics from Columbia, to check his calculation. Bardeen did and suggested that perhaps electrons were being trapped at the surface of the crystal, which would prevent them seeing the calculated effect of amplification.

Next Brattain performed an experiment on the artifact, which confirmed Bardeen's explanation. They were on the right track. They needed to play with the structure of the artifact, trying other configurations. Finally, one configuration worked, but provided only a small amplification. A new kind of electronic valve was almost invented, but the amplification was too small. The group continued to invent alternative configurations. Another configuration placed two closely spaced gold strips on the surface of a dope germanium crystal and an electrical connection on its base. Relative to that base, one gold strip had a positive voltage (called the emitter) and the other a negative voltage (called the collector). They applied a signal between emitter and base. Voila! There was sufficiently large amplification in the voltage between the collector and the base! They called this a 'transistor'. It could do everything the old electron tube could do and would eventually replace it.

Source: (http//:en.wikipidia.org, transistor, 2009)

Figure 2.3. Replica of First Bell Labs Transistor.

AT&T acquired a basic patent for the transistor and made transistors for its own use. AT&T licensed others to make them for other uses. At that time, AT&T was a legal telephone monopoly in the U.S. and legally prevented from entering other businesses. The electron vacuum tube had begun the first age of electronics. The transistor began the second age of electronics. Shockley, Brattain and Bardeen were awarded the Nobel Prize in physics in 1956.

Here is one puzzle about innovation as invention plus commercialization. Although the U.S. innovated the transistor in the 1950s, over the second half century, U.S. firms would lose their dominance in consumer electronics – still inventive but not commercially dominant over the long term. Back in the 1950s, U.S. companies were both inventive and commercialized their inventions. But as time went on in the second half of the twentieth century, many large U.S. firms stumbled. For example Bell Labs, which invented the transistor, was the R&D lab of AT&T. AT&T would be broken down in 1984 into only a long-distance phone service, spinning off regional phone service. In 1996, Western Electric, the hardware manufacturer in AT&T, was spun off as a separate company called Lucent. In 2006, Lucent was merged with the French company Alcatel, becoming Alcatel-Lucent. Bell Labs became the R&D lab for Alcatel-Lucent, still with headquarters in Murray Hill, New Jersey, U.S.A. and with R&D facilities around the world.

History – Invention of the Silicon Transistor

But the germanium transistor suffered from a serious technical performance problem, sensitivity to temperature. What good would transistorized electronics be if their circuit performance changed dramatically from the heat of day to the cool of night, from the hot of summer to the cold of winter? After the original invention of the germanium transistor, electronic engineers yearned for a more reliable and less temperature-sensitive version of the transistor. The obvious technical way was to try to make a transistor not from germanium but from its sister element, silicon.

One of the groups looking for a silicon version of the transistor was based in a then small U.S. company, Texas Instruments (TI). In 1952, Patrick Haggerty was president of TI, and the company was a maker of seismographic detection instruments, sold to oil companies for use in oil exploration. Their instruments then used electron vacuum tubes but needed to be portable and rugged and use little power. It was obvious to Haggerty that transistors would be a desirable replacement for tubes in TI's products. Haggerty hired a physicist from Bell Labs, Gordon Teal. Teal had been researching silicon. Silicon was brittle and difficult to purify as a material for making transistors. Haggerty told Teal and another researcher, Willis Adcock (a physical chemist) to develop a silicon transistor. Many other research groups were seeking the silicon transistor; and it was not an easy artifact to make. But Teal and Adcock did it.

In May 1954, Teal took their new silicon transistor to a professional conference where he listened to several speakers tell of their difficulties in trying to make a silicon transistor. What he heard made Teal happy for as yet no one else had succeeded. When Teal's time to speak came, he stood before the group and announced: "Our company now has two types of silicon transistors in production....I just happen to have some here in my coat pocket" (Reid, 1985, p. 37). Teal's assistant came onto the speaker's stage carrying a record player. This player used a germanium transistor in the electrical circuit for the amplifier. The tiny germanium transistor had been wired visibly outside the record player with long leads. Teal plugged in the player, put on a record and started it playing music. Next the assistant brought onto the stage a pot of hot oil and set on the table beside the record player. Teal picked upon the connected germanium transistor and dramatically dunked it into the hot oil. Immediately, the music stopped as the germanium transistor failed in the oil's hot temperature. Then Teal picked up one of the silicon transistors which he had earlier taken from his pocket and placed on the demonstration table. Teal took a soldering iron (which the assistant had also supplied) and replaced the germanium transistor with the new silicon transistor. The music from the record player sounded again. Then Teal picked up the silicon transistor and dunked it into the pot of hot oil, as previously he had with the germanium transistor. The music did not stop. It continued. The silicon transistor could stand the heat. The meeting exploded in excitement. Texas Instruments had done it! Finally there was a useful transistor – the silicon transistor!

HISTORY – INVENTION OF THE INTEGRATED CIRCUIT (IC) SEMICONDUCTING CHIP

While the silicon transistors were useful, a new problem arose in electronics. The new complex circuits engineers then could dream up might require so many transistors that no one could physically wire them together with the technology of soldering transistors into printed circuit boards. So a new natural limit was on the horizon for transistorized electronics – how many could be physically wired together. In 1958, this was on the mind of two researchers: Jack Kilby at Texas Instruments and Robert Noyce at Fairchild.

Jack St. Clair Kilby graduated as an electrical engineer, and went to work at a firm named Centralab, located in Milwaukee. One evening he attended a lecture at nearby Marquette University given by John Bardeen, one of the inventors of the transistor. The new technology astonished Kilby, and he began reading everything he could find on the solid-state artifice. In 1951, Bell Labs announced the licenses, and Centralab purchased one for the $25,000 fee. They sent Kilby to Bell Labs for a five-day course in semiconductors. From 1951 to 1958, Kilby designed transistorized circuits for Centralab. As these circuits increased in complexity and numbers of transistors, Kilby saw the need for technical breakthrough beyond hand-wiring transistors together.

But Centralab did not have the research capability which Kilby would need to tackle the problem. In 1958, he sent his resume to several larger firms, one of which went to Willis Adcock of Texas Instruments. Adcock hired Kilby, since TI was also worried about the integration problem of transistors. Adcock had a 'micro-module' research project going at TI to approach that problem and to which he assigned Kilby. But Kilby didn't like their approach. One thing was clear to Kilby, since TI led in silicon technology, a good solution to transistor integration should use silicon technology – if it could. So he had this idea. If transistors were made of silicon, could the other components of a circuit (such as resistors and capacitors) also be fabricated on silicon? Why not make the whole circuit on a silicon chip?

It was an inventive idea, for then the technology for fabricating resistors used carbon as the phenomenal material and plastics and metal foil for capacitors. On July 24, 1958, Kilby wrote down in his lab notebook, what he called a 'monolithic idea': "The following circuit elements could be made on a single slice (of silicon): resistors, capacitors, distributed capacitors, transistor" (Reid, 1985, p 38). He entered rough sketches of how to make each component by properly arranging silicon material. Kilby then showed his notebook to Adcock, who said if he could demonstrate the idea worked, an R&D project to develop it would be authorized. Kilby carved out a resistor on one silicon chip and then carved a capacitor on another and wired the two chips together. They worked, showing resistance and capacitance in the circuit. Adcock gave his approval to build a whole circuit on a chip and Kilby chose to build an oscillator circuit – one that could generate a sinusoidal signal.

On September 12, 1958, Kilby was ready. A group of TI executives assembled in the lab for a demonstration. The circuit on the chip was hooked to an oscilloscope which would display the form of the signal from the circuit. Before the circuit was turned on, the Oscilloscope displayed a flat line indicating no signal output. Kilby switched on the new integrated circuit on the silicon chip and it worked – a wiggling line appeared on the screen, the sine-wave form of an oscillator circuit: "Then everybody broke into broad smiles. A new era in electronics had been born" (Reid, 1985, p 39).

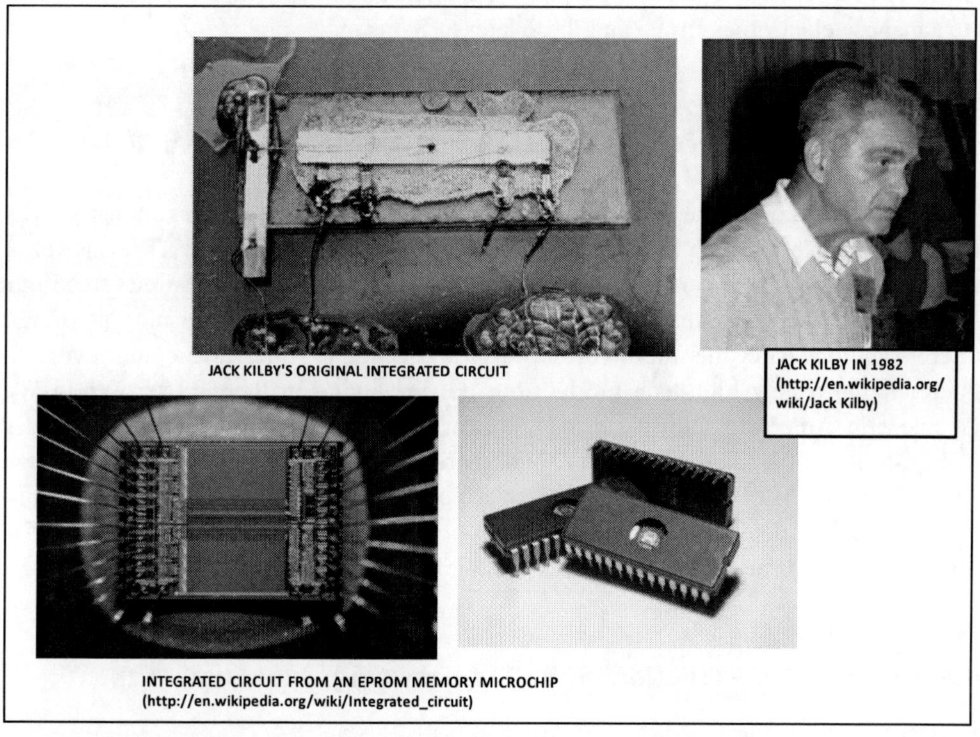

Figure 2.4. Integrated Circuit.

Sometimes a basic invention is made independently about the same time by another inventive mind aware of the technological needs and opportunities. This happened with the invention of the semiconductor integrated circuit (IC) chip. That same summer in 1958, Robert Noyce, who was a physicist and then president of Fairchild Semiconductor, was also worrying about the problem of transistor integration. Fairchild was a new company to manufacture transistors begun by one of the inventors of the transistor, Schockley, after he left AT&T. That following winter on January 23, 1959, Noyce conceived a solution and wrote down in his lab notebook his idea: "It would be desirable to make multiple devices on a single piece of silicon, in order to be able to make interconnections between devices as part of the manufacturing process..." (Reid, 1985, p 41).

Both Texas Instruments and Fairchild Semiconductors filed for the patent on the IC chip. Years of legal argument followed about who had the right to the patent. In 1966, several semiconductor chip manufacturing firms met together; and TI and Fairchild agreed to grant licenses to each other and share in inventing the device. Later, Kilby and Noyce both received (in addition to many other prizes) the National Medal of Science for inventing the semiconductor IC chip.

The IC chip next became the key electronic components for progress in computers. Two kinds of IC chips were designed for computers, logic chips and memory chips. A logic chip contained all the circuits needed to do electronic computation in a computer, called a Central Processor (CPU) chip. A memory chip contained all the circuits needed to store information in memory, called a Dynamic Random Access Memory (DRAM) chip. The two chips were the key components of a computer. Intel grew up to dominate the global market for CPU chips. Other new American firms began the production of DRAM chips but lost their business first to Japanese electronics firms, and these later to Samsung.

INNOVATION = INVENTION + COMMERCIALIZATION

Here is the economic question. If Americans invented the electronics industry, why did American firms lose so much to Asian firms – Japanese, Korean, Chinese? The answer is that innovation consists of two parts, first invention and second commercialization (creating ideas and making products). Americans were great at invention, but not at long-term commercialization. After the middle of the twentieth century, Americans could invest in first commercializing a new high-tech product but not in staying in there. After World War II, management in American industry lost long-term commercial vision, focusing increasingly upon short-term profits. The American capital market lost 'staying power' – lost the desire to invest in the long-term of the U.S. commercial power. To examine the reasons for this, we look the two Schumpeterian technological discontinuities in IC memory chips – in 1981 by Japanese electronic firms and in 2000 by Samsung Electronics.

History: Creative Destruction in the Memory Chip Industry – 1981

The silicon transistor was an amazing invention by Americans. But here is the wonder about American commercial activities in consumer electronics. They did not innovate (commercialize) the silicon transistor for U.S. consumer products. The American invention of the silicon transistor was rapidly used by American firms for business and military markets, but not for consumer markets. This left an opening for another nation's industries to enter the global market of consumer electronics.

Japan took the opportunity. Although through the 1960s and 1970s, new U.S. firms were established to produce IC chips (and dominated the world market). Yet it was the Japanese electronics firms who emerged in the 1980s as the world leaders in consumer electronics. Sony and other Japanese firms innovated the transistorized pocket radio, transistorized television sets, transistorized hi-fi, and home VCR's. This technological substitution of transistors for tubes (technology discontinuity) in consumer electronics products provided the commercial opportunity for the growth of Japanese electronics firms, when U.S. consumer electronics firms lagged behind and were slow to make this change.

The extraordinary success in consumer electronics encouraged the Japanese government to assist industry to go from silicon transistors to silicon integrated circuits, (IC chips). The government wanted to assist Japanese industry to advance from transistors to semiconductor IC technology (another technology discontinuity). By the end of the 1960s, progress in IC chip technology had jumped from putting hundreds of transistors on a chip to putting thousands of transistors on a chip, then called Middle Scale Integration (MSI). The next technology step after MSI was to put tens-of-thousands of transistors on a chip, then called Large Scale Integration (LSI). In 1970, a Japanese government agency, Ministry of International Trade and Industry (MITI) sponsored a national program for LSI next-generation technology. The LSI scale for which the MITI project aimed was intended to bring Japanese electronics firms into competitiveness with U.S. chip firms. This decision by a nation to enter a new technology at the next-generation of the technology (rather than the current generation) was at the time an unusual decision for a developing economy.

MITI provided the equivalent of $10 million in research grants to Japanese electronics firms from 1970 to 1975. Some of the grants went to electronics firms producing transistors and consumer electronics devices and some to firms making chip-production equipment. Japanese firms engaged in the project registered about 1000 patents on chip technologies. Their focus was on the memory chip using LSI technology capability, 16K memory chip for computers, a sixteen thousand bit storage in a dynamic random access memory (DRAM).

Next in 1976, this 16K memory chip was innovated by U.S. firms and by Japanese firms (assisted by MITI sponsored national LSI project). The demand for the 16K chip by 1979 was so large that U.S. firms could not satisfy the demand; and Japanese firms exported to the U.S., gaining a 20% share of the U.S. market. Also U.S. computer manufacturers were surprised to find that the Japanese 16K memory chips were of higher production quality than U.S. produced memory chips. About 20% of U.S. produced chips were defective, whereas only 10% of Japanese produced chips were defective. By 1980, the Japanese electronics firms were fully competitive with the U.S. in memory chip technology and more advanced in manufacturing quality.

Then in 1980, a next-generation of memory chip was aimed at in the 64K DRAM chip. Making it would require scaling down the size of transistors from LSI processing technology to Very Large Scale Integration (VLSI) technology. In 1981, the race was on between U.S. and Japan‘s chip makers to the first to innovate (commercialize) the 64K memory chip. The first to the market would likely grab the majority of the market. Japanese firms were the first to market in November 1981.

U.S. firms had designed a 64K chip with some refined features over the 16K chip design to produce a smaller sized chip, but this redesign slowed down the timing of their product introduction. In contrast, Japanese firms took a standard 16K chip design (innovated earlier by the U.S. firm Mostek) and simply enlarged the design to the 64K level. They understood the importance of being the first to market. This resulted in a larger sized chip but at the time made little difference to the computer customer. What mattered to the customer was speed (who produced the chip first) and quality (fewest defective chips). In September of 1982, U.S. chip makers introduced their 64K memory chip, but it was too late. Japanese DRAM chips already had 76% of the market. U.S. memory chip-makers died from lack of cash flow in selling their chips. At this technology discontinuity, Japanese industry better understood the conditions for success in global competitiveness than did their U.S. counterparts.

The American semiconductor memory chip firms never recovered. The U.S. chip industry stopped producing memory chips (with a few exceptions). When chip density increased by an order of magnitude in density, the cost of building a chip processing facility to produce that density also increased by about an order of magnitude. Unless a chip producer had a substantial market share, there was not the cash flow to finance the investment for the next-generation production facility. When the U.S. chip producing firms lost market share, they lost the cash flow to finance the development of the next-generation of technological production capability. Once out of the race for memory chip markets, U.S. firms were out of that race for good.

Theory: Moore's Law as the Technology Roadmap for the Chip Industry

How had all this technology planning occurred in the global IC chip industry? There was industrial planning for MSI (middle scale integration) and then LSI (large scale integration) and then VLSI (very large scale integration). This international industrial plan happened because everybody knew the technology roadmap for progress in IC chip technology – everybody. That knowledge was out there in the world, because back in 1965, Gordon Moore had told everybody.

In 1965, Gordon Moore was the director of the R&D Lab of Fairchild Semiconductor, and he was continuing progress in IC chips. Moore wrote: "The complexity (of IC circuits) has increased at a rate of roughly a factor of two per year. Certainly over the short term this rate can be expected to continue, if not to increase. Over the longer term, the rate of increase is a bit more uncertain, although there is no reason to believe the rate will not remain nearly constant for at least 10 years (doubling each year)." (Moore, 1965). And so it happened, over the next twenty years, as shown in Figure 2.5.

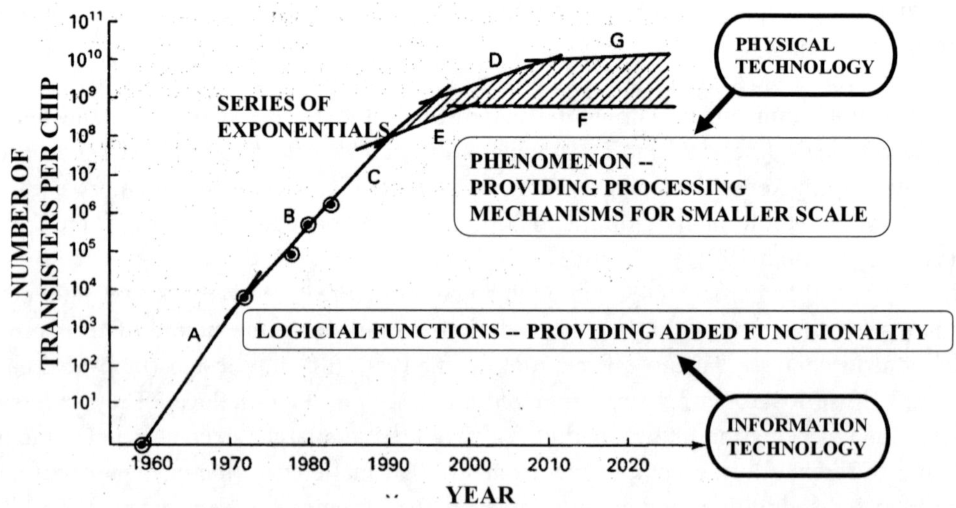

Figure 2.5. Technology Progress in Semiconductor Chip Transistor Density – As Forecasted In 1983.

Moore's law was not a law of 'nature'; it was an industrial plan. The semiconductor industry planned to make it happen. Moore's law was used by industry, globally, as a guide to long-term technology planning, a *technology roadmap.* One notes that in Figure 2.5, the scale is semi-log, so the progress from 1960 to 1985 had been in two exponentials curves (graphed as straight-lines in semi-log charts. Ever smaller transistors were fabricated with new processing technologies. This enabled the rapid increase in numbers of transistors on a chip. It was progress in chip-processing technologies which made possible the progress in IC chips. The increasing density in the transistors on a chip made possible the vastly improved logical functions of processing information by the chip. *Information processing by IC chips is the 'function' of the chip technology; while transistor configurations of the chip are the form (morphology) of the chip technology – function and form in technology.*

The limit to Moore's law would occur when the physical features, in the constructing a transistor out of silicon atoms, would get down to nearly the size of a silicon atom itself. So the technology forecast even back in 1985 was for continuing exponential progress until after 2010. Then the 'natural limit' of progress in the technology would be reached as the chip-making-technology approached features nearly the size of a silicon atom. The natural limit is the transistor feature size, of about hundred silicon atoms (or 10 nanometers). In 2012, chips were being produced at about the 20 nanometer scale and were expected to approach the 10 nanometer size around 2020. Thereafter, there would be no more Moore's law. The amazing historical thing is that this technology foresight by the chip industry in 1985 has been reasonably accurate. They knew then (as now) that silicon atoms are finite in size. Nature eventually limits all technological progress, providing a natural limit for all physical technologies. *This is because all technologies are ways to manipulate (alter, use) nature; and limitations of how much one can manipulate nature limits technological progress.*

The IC semiconductor chip industry is an industry in the history of the industrial revolution which progressed by a sequence of technology discontinuities (next-generation technology progress). Industry planned progress as technical discontinuities, using Moore's law as a plan for action. As a plan, Moore's law enabled Schumpeterian 'creative destructions' in the industry in the years of 1981, 2000, and 2008.

Moore's law had provided the technology roadmap to the world. But which companies had the financial power and commitment to carry it out? This would determine who captured the market. In 1981, at the technology discontinuity of the 64K Dynamic Random Access Memory (DRAM) chip, Japanese firms drove U.S. firms from the DRAM market. In 2000, Korean firms drove Japanese firms from dominance in the DRAM market. And in 2008, the Korean firms also drove Taiwanese firms from the market.

Theory: Creative Destruction in the DRAM Industry in 1981

Technology progress in IC chips had been anticipated and planned according to Moore's law. But which firms benefited from this futurist knowledge? It was the firms willing and capable of making the large investments of capital necessary, to take practical advantage of the knowledge about the future. The move from a technology discontinuity to an economic discontinuity is what Schumpeter had called 'creative destruction' in the dynamics of technology and economy. *In creative destruction it is planning, commitment, and capital which make a difference in capturing the benefit of knowledge about a technology's future.* In

Figure 2.6, we describe the theory of the industrial life cycle curve for the DRAM industry as it actually occurred in history from 1961 to 1981 (with the addition of the 16K and 64K disconinuties.)

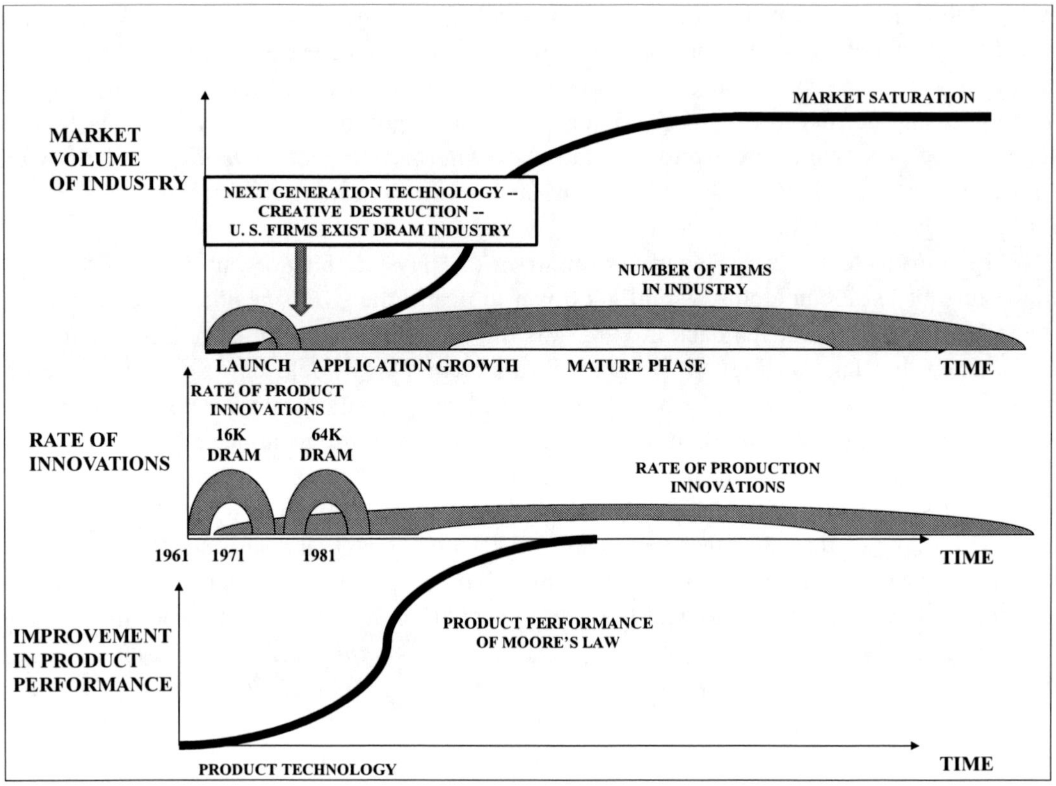

Figure 2.6. Early Dynamics of Dram Memory Chip Industry.

Due to the planned technology progress by the semiconductor IC industry in the 1960s and 1970s, the product performance for the DRAM memory chips did not level out but continued to jump in 1970 and 1980 as the technology discontinuities of the next-generation products of 16K and 64K DRAM chips. But the market capture of the 64K DRAM chip by Japanese firms in 1981 (eight months before the U.S. firms introduced their 64K chip) had put the U.S. firms in jeopardy. Without an immediate cash flow from their late-introduced 64K chips, the U.S. memory chip firms could not afford to continue in business (and compete for the next generation of DRAM chip at 264K). Market capture of the memory chip industry by the Japanese firms moved the industry from the U.S. to Japan – Schumpeterian creative-destruction.

In the case of IC chips we now see that additional concepts must be added to tradtitional innovation theory. The original key ideas of an industrial-life-cycle were: *(1) the rate-of-innovation in the core-technology of the product, (2) the rate-of-innovation in the core-technologies of its production, (3) number of firms entering and existing in the industry.* Now in the memory-chip industry, we should add the additional key ideas of: *(4) next-generation technologies and (5) creative destruction.*

History: Samsung's Creative Destruction in the Memory Chip Industry – 2000

In 1983, this knowledge of Moore's law and the event of Japan's creative destruction of the American memory chip industry were two notions upon Samsung would build its strategy, for entering semiconductor memory chip industry. Tony Michell wrote: "Why did Samsung Electronics . . . continue to invest money in (electronic) components. The answer in this strategy lay at the heart of what it meant to be Korean in the . . early 1980s. Despite escaping from Japanese colonization in 1945, Korean industry remained heavily dependent on Japanese know-how and even more on Japanese components. . . Unless they could penetrate into the component business, Korean industry would forever be dependent on the Japanese, who were now dominating the electronics industry. . . Chairman Lee in February 1983 announced that Samsung would make full inroads into the semiconductor industry." (Michell, 2012, p. 22)

It took two more years, from 1983 to 1985, before Samsung produced its 64K DRAM chip. This was four years after their Japanese competitors had produced their 64K chip in 1981. There was a lot of catching up to do before Korea would be competitive with Japan in IC chips. Samsung was still 'path-following' and needed to go to 'path-revealing' before they could triumph with 'path-creating' and demolish the Japanese memory chip industry. But they accomplished all this, as is summarized in Figure 2.7.

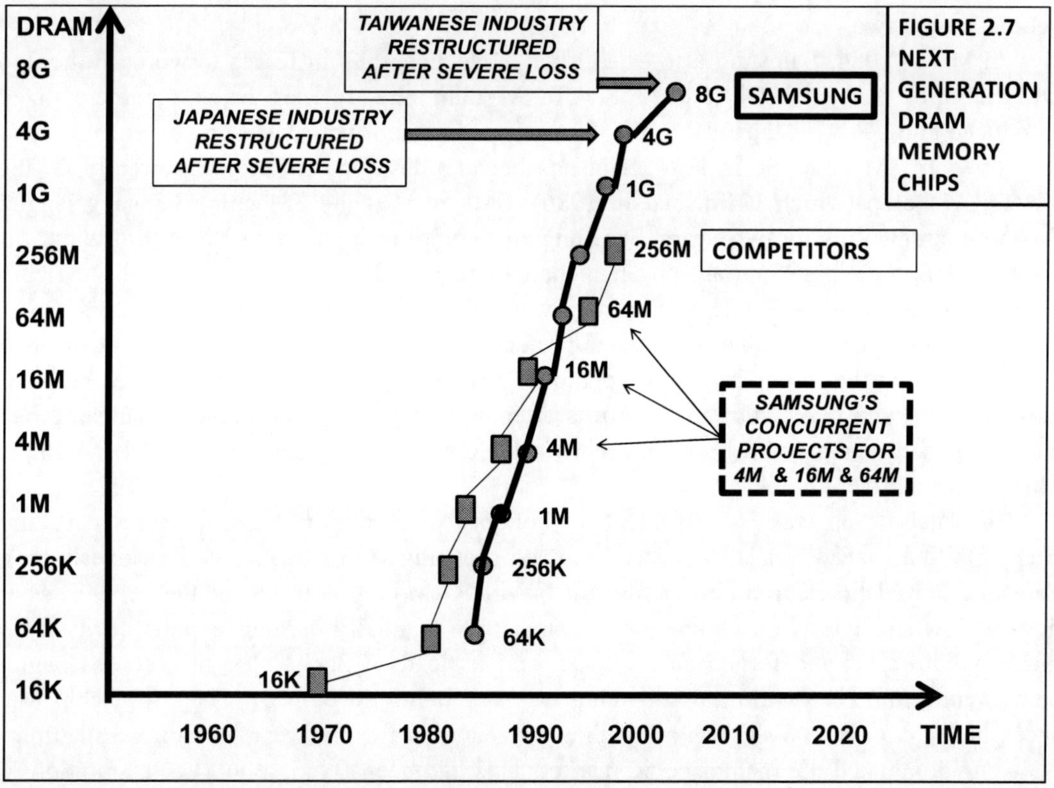

Figure 2.7. Next Generation Dram Memory Chips.

In 1985, Samsung‘s 64K DRAM memory chip was late in the world market. The global IC chip technology race had already innovated the 256K chip, and the 1M chip was being developed. Samsung had to catch up in the technology race with international competitors before it could make substantial profits on its investment. We recall that in 1981, the U.S. chip makers had been forced to drop out of the race, at the 64K technology discontinuity. No longer were they making the magnitude of sales and revenue to be able to afford investing in the next-generation-technology chip of 256K.

Why was Samsung able to enter the race late and stay in and eventually to win? It was because the Samsung group had both the determination and capital for the race. The Samsung Group was a Korean chaebol. It could finance the continuing development of memory chips, not from selling memory chips, but from selling other products through its chaebol member companies. The structure of capital in Korean chaebols was important to Korean firms' capability in continuing to invest in manufacturing development for the long duration of the chip race – up along Moore's law.

Within a chaebol capital structure, Samsung Electronics began development of the VLSI stage of the 64K DRAM chip in May 1983. It contracted with the U.S. firm of Micron Technology for the chip design and with a Japanese firm for the production technology. By November 1983, six months later, the Samsung succeeded in its development. But it was still two years behind Japanese producers. So in September 1983, just as the Korean group was finishing the 64K chip, a Samsung group in America began designing a 246K DRAM chip. Samsung was concurrently developing two generations of DRAM chips in 1983, not one generation but two.

In March 1984, Samsung also established a second group in Korea to work on the 246 DRAM chip. Yet in 1984, the 256K DRAM chip already was being commercialized (produced and sold) by Japanese firms. Samsung would not yet 'catch-up' in technology at the 256K DRAM stage. So in 1984, Samsung began a development project currently for the 1M DRAM chip (which it finished in 1986). Still even at the 1M chip, Samsung lagged Japanese competitors by two years. Samsung still could not make substantial profits until it caught up. So late in 1986, Samsung launched a concurrent project for the 4M DRAM chip (accomplishing it in 1988). Then Samsung was only a half year behind competitors. So in 1987, Samsung started another concurrent project to develop the 16M chip, which it finished in 1989 – but at the same time as competitors. Now Samsung could export DRAM chips of 16M capacity and make substantial profits after its five years of investment. Samsung had caught up with competition by concurrently developing three generations of DRAM chips: 1M and 4M and 16M chips.

But catching up was not sufficient to 'conquer' competition. Samsung wanted to and needed to 'leap ahead'. In 1989 (instead of just producing 16M chips to recover investment), Samsung started a concurrent project for the 64M DRAM chip. It produced this in 1992 – but now with a small lead on competition! Samsung was finally earning export profits from memory chip sales and forged ahead, to expand its lead. A 256M chip project was begun concurrently in 1991 (while the 16M chip was still being developed). The 256M chip was successfully commercial in 1994; and then Samsung had a one year lead on competitors! Samsung dominated the memory chip market. This had been Byung-chull Lee's ambition in 1983; and his son, Kun-hee Lee, achieved memory chip dominance in 1994.

Samsung pressed on, continuing the race. In 1993 as the 256M chip was still being developed, Samsung had a second team starting the 1G chip. In 1996, Samsung innovated the

1G chip ahead of competitors by one and a half years. Samsung had begun the 4G chip development concurrently in 1993 and commercialized it in 2001.

All those generations of DRAM chips (16M, 256M, 1G, 4G) had used millions of dollars to establish production capacities. Now in the lead in 2010, Samsung earned cash flow to earn back all its investment and much more. Lagging competitors didn't earn profits. In 2000, Japanese firms (which had twenty years earlier seized the memory chip business from the U.S.) quit the memory chip business after severe financial losses. Korea had conquered in the Moore's Law race of technology discontinuities. And in so doing, Samsung created a Schumpeterian creative-destruction of the Japanese DRAM chip industry.

Samsung went on in the 2000s to innovate the 8G and 16G and 32G chips. Then Taiwanese competitors dropped out. By 2005, Samsung dominated the DRAM memory business by 32% and the NAND flash memory business by 51%. In 2010, Samsung earned 10 billion dollars profit alone from its semiconductors. Each year, Samsung was annually earning several times its investment to capture the market.

LEE BYUNG-CHUL (LEFT), SAMSUNG GROUP'S FOUNDER, AND LEE KUN-HEE (SECOND FROM THE LEFT), FORMER CHAIRMAN OF SAMSUNG GROUP, ARE HAVING A MEETING WITH OTHER BOARD MEMBERS IN 1978. (HTTP://WWW.KOREAITTIMES.COM/STORY/7223/ HAIL-FATHER-BUSINESS-LEE-BYUNG-CHUL)

Figure 2.8. Samsung Group Founder.

Theory: Long-Term of the Chip Memory Industrial-Life-Cycle

To explain the industrial life-cycle for IC chips, one need to add to innovation theory. For example, one can see in Figure 2.6, that the theory did explain industrial dynamics for an industry with only a single radical innovation in its core-technology, which was the gasoline/diesel internal combustion engine, engines which still drive cars and trucks, Figure 2.9.

But this pattern did not occur in the IC chip industry, and one needs to modify that theoretical pattern, as shown in Figure 2.10.

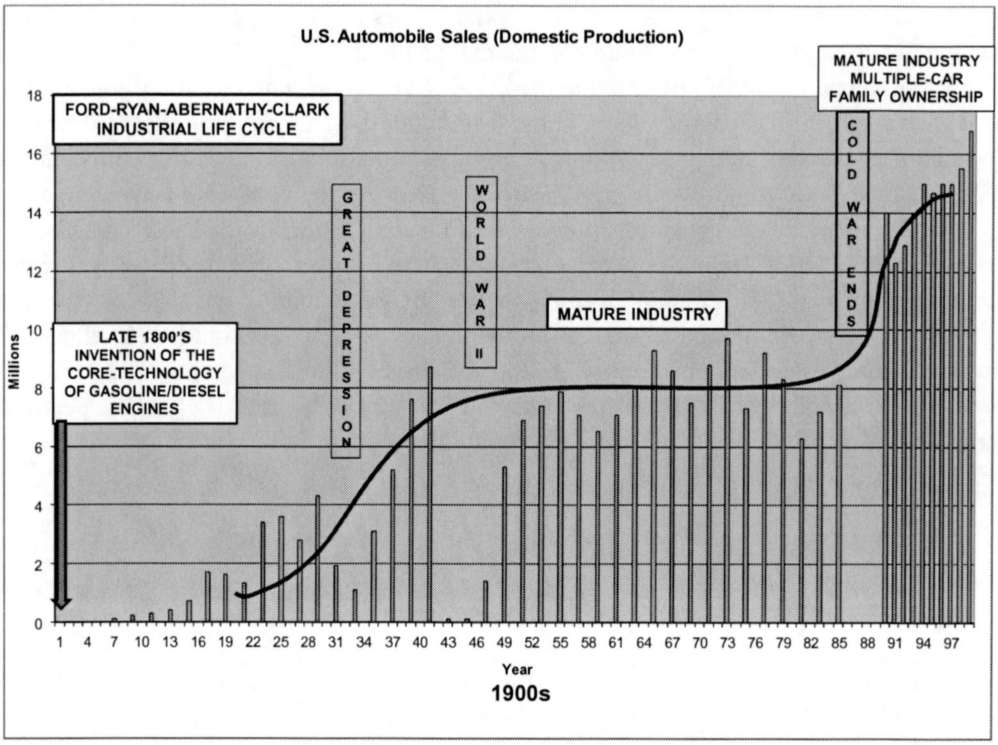

Figure 2.9. Us Automobile Sales (Domestic Production).

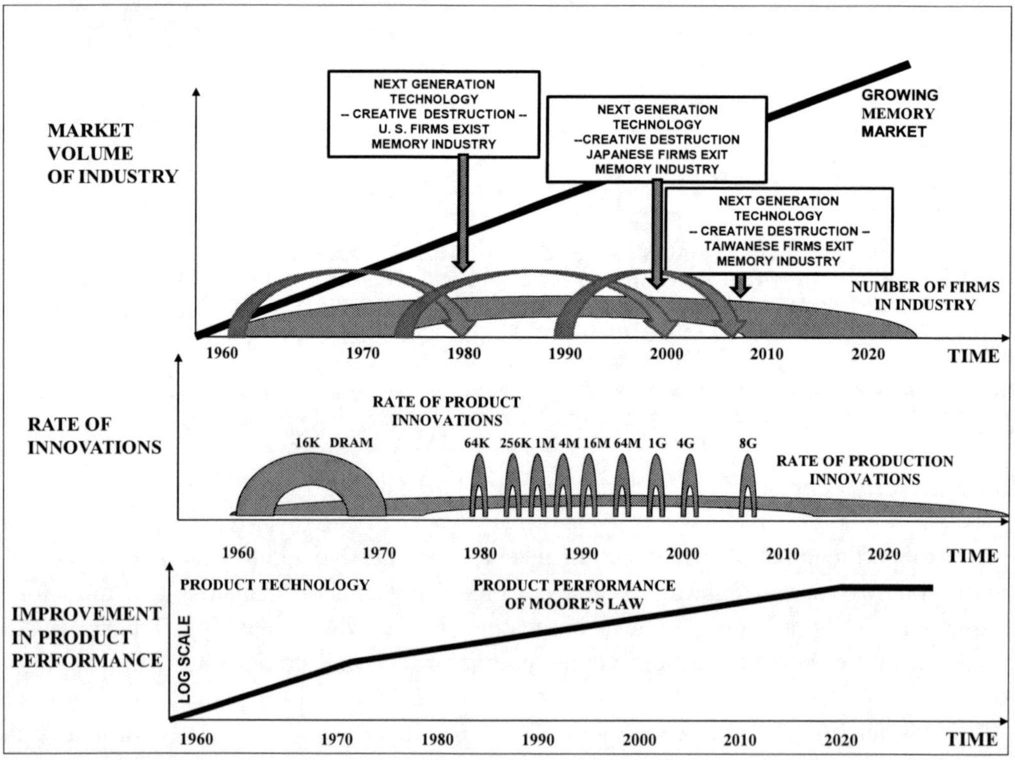

Figure 2.10. Long-Term Dynamics of Dram Memory Chip Industry.

In a multiple next-generation technology history of a product, its industry has a very different form from that of an industry with only a single radical innovation. There is no single peak performance of a product, but several generations of increasing performance. Production innovation is required at each generation to produce new products. And creative destruction of an industry can occur at a technology discontinuity.

Product performance did not peak in the early days of the chip industry but instead continued to improve in performance due to next-generation chips. The next-generation chips were enabled by next-generations of production technologies. It was these technology discontinuities in production techniques which enabled the ever smaller transistors to be produced which in greater densities for increased memory capacity.

As the product performance was continuing to exponentially improve, the applications of the products expanded, and the market for memory chips did not saturate. Market growth continued to increase over the long-term of the industry from 1960 through 2020.

But at each technology generation, the cost of factories increased by an order of magnitude. Progress in IC technologies came at ever larger capital costs. This is why speed-to-market in a next technology generation was so important. The first to a technology-discontinuity market captured the sales revenue, enabling it to continue R&D investment and factory investment for the next-generation-technology. When a firm fell behind in speed-to-market, then sales ended and funds for continuing investments stopped. The firm went out of the business of memory chips. But the Schumpeterian 'creative destruction' happened not just to an individual firm but to a nation's firms – national industrial destruction. Why? The answer lies in the capital structure of a nation for industrial development.

History: Global Capital Structures after World War II

The capital structures for the U.S., Japan, Korea, and Taiwan were all different at the times of creative destruction in the memory chip industries. At the time of the 1981-82 creative-destruction of the U.S. memory chip industry, the U.S. interest rates (in the U.S. capital structure) on commercial loans ranged from 6% to 9%. This compared to 3% interest which the people of Japan were receiving from savings accounts in the Japanese Postal Office, and these savings were available to the keiretsu as a source of capital. This meant that capital costs for the Japanese chip industry were one-half to one-third less than capital costs for the U.S. chip industry – putting the U.S. firms at a serious disadvantage in manufacturing investment.

A keiretsu is a group of interlocking companies controlled by a core bank. It can facilitate innovation by the mutual assistance companies can technically provide within their keiretsu and by the financing which the core bank can provide for the commercialization of new technologies by its companies. This form developed early in the industrialization of Japan and was restored during Japan's post-war reindustrialization in the 1950s. Also by 1980, Japanese electronic firms produced not only chips but also consumer electronic products, generating revenue from both components and consumer products. Also earlier in 1971, they had been assisted by the Japanese governmental agency MITI to acquire the semiconductor technology of Large Scale Integration, producing 16K memory chips by 1976. These electronic firms were each a part of a keiretsu and financed by their keiretsu bank to build 16K chip production capacity in 1976 and then, significantly, 64K chip production in 1981.

In contrast, the U.S. semiconducting chip firms were small single business firms, and did not produce electronic consumer goods nor have a cash flow from any product other than their semiconductor IC chips. They belonged to no corporate group, with a bank at the center, such as a keiretsu. Consequently, when keiretsu-financed Japanese electronic firms grabbed the 64K DRAM market first, U.S. firms had no cash flow to continue nor to invest for the next generation of chips. All U.S. firms singly devoted to memory chips went bankrupt. The only U.S. semiconducting IC firms to survive were those which produced logic chips (e.g. Intel, Motorola, Texas Instruments). The Schumpeterian creative destruction of the U.S. national memory chip business was due to capital structure and corporate organization in the U.S. – which together had not facilitated investments in next-generation chip product.

The next Schumpeterian creative destructions occurred in 2000 for the Japanese memory chip industry and in 2006 for the Taiwanese memory chip industry. Again, respective capital structures played an important role in those technology discontinuities. Samsung was a commercial group called a 'chaebol'. The chaebols were groups of interlocking businesses set up or acquired by an individual, sort of a 'family empire'. Chaebol companies owned stock in each other and were able to finance each other's operations, because all ultimately were under control of the founding family. In the early industrialization of Korea after 1961, government policy encouraged the chaebol to acquire foreign technology and build new technology-based businesses to export Korean products. Then government selected some chaebols to undertake major technology acquisition and development projects and guaranteed their foreign loans. In the Samsung Group, cash flow from other Samsung companies and foreign loans provided the capital for Samsung Electronics to succeed in the development and production of the series of technology discontinuities from the 64K DRAM chip to the 4G DRAM chip. In 1996-97 during the Asian financial crisis, South Korean was hard hit in this crisis. The Samsung group had large foreign debts, but it used the crisis to restructure the group onto a sound financial basis. Afterwards, with large revenues from the 4G and 8G DRAM and NAND flash-drive memory discontinuities, Samsung dominated memory chip production. The Korean Schumpeterian creative-destruction drove first the Japanese and then the Taiwanese competitors out of the memory chip business.

Theory: Long-Term Innovation as (Invention + Commercialization + Market Capture)

Management of technology had from its inception defined technological innovation as invention + commercialization. (Betz 1987) But now one can look back and see that this definition is principally for the early-history of an innovation. Over the longer history of a technology innovation, the firms which had been the 'first movers' in the early days rarely continue to dominate over the long term. For example, Ford Motor Company was the first mover in the U.S. automobile industry but lost its industrial dominance to General Motors after the 1930s. General Motors lost its U.S. dominance to Toyota in the 2000s (with GM even going bankrupt in 2008). In computers, IBM dominated from the 1950s to the 1970s, but lost dominance after the 1980s. In xerography, Xerox innovated and dominated until the 1980s but then lost to Japanese firms. In the semiconductor memory industry, market dominance passed from U.S. firms to Japanese firms to Korean firms.

Accordingly, over the long-term history of an innovation, the concept of 'technological innovation' needs to include one more idea: invention + commercialization + market capture. Initial commercialization by an innovating firm can capture a market, at first. But longer-term market domination may be captured later by other firms. If later market-capture involves destruction of a national industry, then a 'national market-capture' introduces Schumperterian creative-destruction on a global scale.

SUMMARY: CONCURRENT DEVELOPMENT PROGRAM MANAGEMENT IN SAMSUNG

The technology roadmap for the sequence of next-generation-technology memory products was globally available, expressed clearly in Moore's law. Samsung used this in its plan to launch technology/product development in concurrent programs for next-generation-technologies (NGT):

- The 64K DRAM chip development program was launched in May 1983 and went into mass production in September 1984. But during that time, a second concurrent 256K development program had been launched in September 1983 (four months after the 64K launch). So the 64K and 256K development projects had run concurrently in Samsung Electronics.
- The 256K DRAM chip went into mass production in Samsung in April 1986. But a 1M development program had been launched in May 1985, running concurrently along with the completion of the 256K program.
- The 1M DRAM chip went into mass production in November 1987. But a 4M development program had been launched in January 1986. So for a time, three NGT memory chip programs had been run currently: (256K from Sept 83 to April 86; 1M from May 1985 to November 1987, and 4M from January 1986 to November 1989). So in 1986, three NGT programs had been running concurrently.
- The 4M DRAM chip went into mass production in 1992, having been concurrently launched as a development program in 1989 during the finish of the 4M chip program.
- The 256M DRAM chip went into production 1994, having been concurrently launched in 1991, alongside the 4M program.
- The 1G DRAM chip went into production in 1996, having been concurrently launched in 1993, alongside the 256M program.
- The 4G DRAM chip went into production in 2001, having been launched in 1998 (two years after the finish of the 1G program (Choi, 2007)

Samsung had run several NGT programs concurrently to catch-up and pass the competition in the semiconducting chip industry. Samsung technology strategy was as 'Path-Following', 'Path-Revealing' and 'Path-Creating'.

For the 64K and 256 DRAM chip programs, Samsung used the 'Path-Following' phase, acquiring chip design technology from the U.S. and production technology from Japan. For the 1M, 4M and 16M, Samsung had used the 'Path-Revealing' phase, focusing upon the problem-solving for denser transistor chip production technologies; and in this phase, Samsung had caught up with its competitors in 1989. Then for the successive NGT chips, Samsung had used the 'Path-Creating' phase, focusing upon originality in design and production to leap ahead of competitors. The use of concurrent development programs allowed Samsung to speed up the technology discontinuities to commercialize innovative NGT chips ahead of competitors. It was this speed-of-innovation enabled by development-program-concurrency which allowed Samsung to capture the memory chip market and destroy competition.

Lessons Learned

In trying to understand the reasons for Samsung's amazing success in the global market of memory chips in the twenty-first century, we used several important concepts:

- Technological Innovation = Invention + Commercialization + Market-capture,
- Importance of Capital Structures to Innovation,
- A revised Industrial-Life-Cycle for technology discontinuities and creative-destructions,
- Technology Roadmap for Next Gerneration Technology (NGT) Concurrent Development Programs in a Speed-To-Market.

In the concept of technological innovation, the long-term history of an industry should include the idea of market-capture along with the ideas of invention and commercialization. Technological innovation does depend up 'invention' and 'commercialization' to begin the industry. But which firms dominate and when they dominate is not explained only by these ideas. The additional idea of 'market-capture' is needed to explain competition within an industry over its long technical history.

Capital structure is an important relation to innovation. It is well known that a structure of venture capital is essential to radical innovation in its early stages of commercialization. But what has not been so well studied is the relationship of a capital structure to the later history of innovation, in market-capture. As we have seen in the case of the IC memory industry, the capital structures of keiretsu in Japan and chaebols in Korea played important roles in sustaining the financing of innovation through several technology discontinuities.

Because of the importance of capital to the commercialization, a Next Generation Technology (NGT) discontinuity in the core-technologies of a product offers opportunities for restructuring of an industry – creative destructions.

When a series of NGT discontinuities can be technically planned by a technology roadmap, then concurrency in NGT development programs can be used to speed up the pacing through the technology discontinuities. But it takes a clear NGT technology roadmap, a strong CEO determination, capital-intensive capabilities, and program management skills to successfully carry out concurrent development of NGT products.

From the Samsung case, we can see that success in global completion required innovation capability and also management leadership and access to large capital. These three factors in innovation – technology roadmap and leadership and capital – must together be implemented in a long-term strategy to capture a global market.

> *On a global scale, technology discontinuities provide opportunities for economic creative-destruction – but only for competitors determined and capable of the long-term effort required to accomplish market dominance.*

> *When technological discontinuities happen, market dominance is captured by the company with (1) a technology road-map, (2) a long-term capital commitment, and (3) leadership with strategy and rapid implementation.*

Chapter 3

PRIME MOVER IN GLOBAL COMMODITY INDUSTRIES – POSCO

INTRODUCTION

We looked at how Samsung Electronics leapt ahead of global competition in semiconductor memory chips through rapid innovation of next generation technologies – as electronics was an industry of rapidly changing technology through the twentieth century into the twenty-first century. But technologies are not always changing rapidly. In fact, most industrial technologies are stable. How did South Korean industry succeed in entering and competing in a mature technology industry – an industry producing commodity-like products, such as the steel industry? We next examine the case of POSCO (Pohang Iron and Steel Company), successfully climbing to the fourth largest steel company in 2010, in only forty years since its beginning in 1968.

A commodity industry is one in which core technologies in products have matured, and further technological progress is focused upon production. William Abernathy had introduced the concept of an 'industrial life cycle' to depict the transition of a high-tech industry into a commodity industry. (Abernathy, 1978) This is when the rate of product innovations peak and then declines; and yet the rate of production innovations increase and continue. During this transition the numbers of firms in the new industry also peaks and then declines.

Alfred Chandler addressed the problem of which new firms in the new industry would rise to market dominance and called these 'first movers'. (Chandler, 1990) These are the companies which move first to: (1) continue to advance the new technologies, (2) develop large-scale production capacity, (3) develop a national distribution capability, and (4) develop the management talent to grow the new firm. (Chandler, 1990)

The reason investment in advancing technology is important is that a competing company must not lag behind technologically and can gain competitive advantage from being innovative. Investments in large-scale production capacity and a national distribution system are necessary for an emerging firm to gain a dominant market share in the new national market. Also now in a global economy, a first-mover firm must establish presence in both national and international markets. Developing management talent to run a growing, large firm is also a necessary investment. For example, a frequent kind of failure of many firms to have a long term survival after an initial success has been the failure of the founder of the firm, as an entrepreneur, to build a management team that can succeed the founder. Accordingly, the milestones a new venture must meet to become a first mover are: (1) Technology innovation, (2) Production and distribution expansion, (3) Organizational and management development, and (5) Capital acquisition.

And in new industries, such companies as Ford (automobiles) IBM (computers), there had been first movers which rose to dominance. These two concepts fit historical patterns in the early times of new high-tech industries – but not over longer times. First movers have not always maintained dominance in a market. For example, Ford lost U.S. dominance to General Motors after the 1920s; and General Motors lost dominance to Toyota after the 1990s. Xerox lost dominance in printing in the 1980s; and IBM lost dominance in computers in the 1980s. Why? What are the criteria for market dominance in commodity-type industries?

Leadership in the form of the founding entrepreneur has always been important. For example, in 1864 after the Civil War in the U.S., Andrew Carnegie established a first mover in the steel industry in Carnegie Steel Co. He adopted the new Bessemer process for steel making as a new technology and organized vertical integration of suppliers of raw materials. Carnegie Steel continued to expand by buying competitors. But in 1907, it was merged into U.S. Steel. U.S. Steel was a dominant U.S. producer of steel throughout most of the twentieth century. Yet by 2011, it ranked only 13 in the list of the world's largest steel companies. (World Steel Association, 2012) What happens to leadership in a large company, after innovation slows and industry becomes a commodity industry? Why can first movers fall from industrial dominance in an industry, after it becomes a commodity industry? What are the critical criterions in competition over the long-term of an industry, as a 'commodity' industry? What role does technology and innovation play in a commodity industry? How should technology be managed in a commodity industry?

Technological Innovation and Creative Destruction

We recall that in 1942 Joseph Schumpeter had introduced the term 'creative destruction' to describe how radical technological innovation disrupts industrial structure in technology discontinuities. (Schumpeter, 1942) In 1978, William Abernathy examined innovation in the American automobile industry, pointing out that while product innovations had historically peaked in the industry, innovation in production had continued. Later in 1981 to describe this more clearly, David Ford and Chris Ryan proposed a chart of market-volume-over-time for an industry to reflect the underlying maturation of the core technologies of its product, calling it an 'industrial life cycle' (Ford and Ryan, 1981). William Abernathy and K. B. Clark described how in the early days of this pattern, the product innovations peaked and then were followed by production innovations. (Abernathy and Clark, 1985) Next the industrial life cycle was compared to the product technology S-curve and numbers of industrial competitors in order to show their relationships, Figure 3.1. (Betz, 2011).

The innovation of products based upon a new radical core technology creates a new industry to produce the products. In the launch phase, many new product innovations spur the growth of the new industry. But after a design stand is achieved, the rate of product innovations decline. Innovation shifts to production innovations (producing products at a design standard) for reducing costs and improving quality. As technical innovation both in products and production slows, then the industry enters a mature industry commodity-type phase. All products by competitors are similar in performance and cost; and differences in product design principally differentiates product competition. The number of firms in the new industry increases rapidly in the launch phase to peak in the application growth phase and then decline in the mature industry phase.

This pattern describes the early history of a new industry based upon a radically new core-technology. But it does not accurately describe the later history, after the industry moves from a high-tech-product industry into a commodity-product industry. How does an industrial life cycle appear later in its commodity history? To examine this, we analyze the case history of the Korean steel firm POSCO in the second half of the twentieth century, long after steel had become a commodity industry.

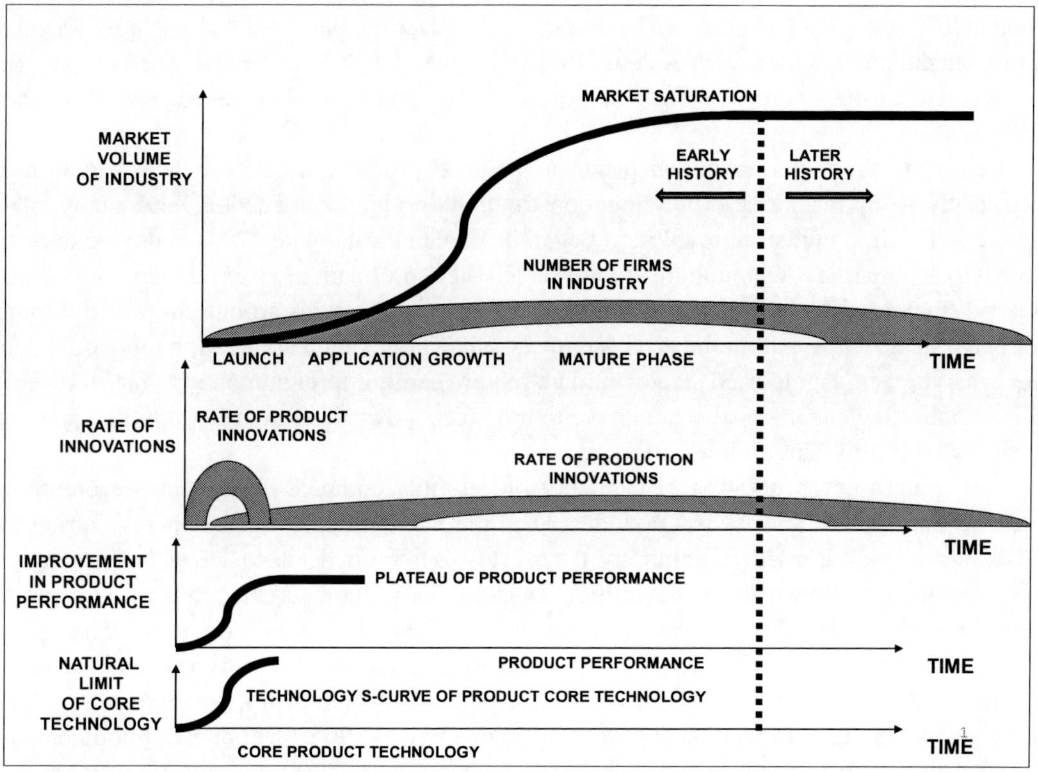

Figure 3.1. Dynamics of New Industries Based on Technological Innovation.

HISTORY OF POSCO

The concepts of the industrial life cycle and first movers do not include the context of an industry. But in real cases of history, one must include the context of industrial dynamics. In the case of POSCO, the context begins with: (1) the division of the liberated country of Korea into two nations, North and South Korea, (2) the Korean war between them, (3) the first republic of South Korea under Syngman Rhee, (4) the second republic of South Korea under

Chan Mon, and (5) the third republic under Park Chung-hee. POSCO began as a government program under President Park Chung-Hee.

After liberation from forty years of Japanese rule in 1946 and then a repelled invasion by North Korean communist government in 1950-53, South Korea was an undeveloped country in 1961 with no industrial base – and with only half the Korean population and still technically at war with North Korea. Park had been a general in the Korean war, and in 1961 led a military coup to take over the ineffective government of the second republic. Under pressure from the U.S., South Korean government was restored to a presidential system; and by a narrow margin, Park won an election for president in 1963.

President Park proposed a five-year economic development plan for an export-oriented industrialization. His slogan at the time was 'Development First, Unification Later'. O Won Chol wrote: "During the era of the export-first principle, 1964-1970, President Park set an annual 40% growth in export as the standard . . . During the era of the export-oriented-industrialization,1973-1980, President Park set the 10 billion dollar mark and The achievement of 40% annual growth in exports as the goals to be achieved." (O Won-chol, 2009)

Central to the industrialization plan was national production of steel. It is widely used industrially – in machinery, building construction, bridge construction, and many other products. The first industrial revolution began in England in the late 1700s and developed on the basis of three new technologies (coke-fueled steel production, steam engine, and textile factory). Inexpensive steel continued to drive the industrializations around the world through railroads, steam ships, automobiles, skyscrapers, etc. Korea's industrialization too would need inexpensive steel (as imported steel would be require earning foreign money to buy the steel). This was the challenge. Korea would have to produce its own steel in order to build an export economy – in ships, automobiles, tools, etc.

But could a new Korean steel company successfully compete in the world economy of steel? Joseph Innace and Abby Dress described the international situation then: "During the 1960s, steel was a growing commodity the world over. From 1960 to 1970, Japan's annual steel production surged from 22 million tons to 93 million as the country's economy expanded . . . . the U.S. steelmaking sector added nearly 50 million tons of capacity. Expansion plans were taking shape in the European Community, and several developing countries outlined ambitions programs . . . . The lack of a steel industry, however, was a thorn in the side of the growing nation (of Korea). From 1961 to 1966, steel demand doubled, and in 1966, alone, Korea imported 350,000 tons compared to 212,000 of domestic production." (Innace and Dress, 1992, p. 32)

Korea was importing over half of its steel needs. In 1965, President Park visited the U.S.: "President Park went to Pittsburgh. . . meeting senior executives of the American steel industry and was treated to a first-hand tour of an integrated steel mill. He liked what he saw." ((Innace and Dress, 1992, p. 33) Park next tried to arrange international financing for construction of an integrated steel mill in Korea. By the fall 1967, he obtained a preliminary agreement with foreign banks and steel companies to provide $95.7 million in capital to build a steel company, along with $35 million from Korean capital. In 1968, President Park Chung-hee gave the assignment to build a new steel plant to his younger colleague, Park Tae-joon, who would build the Pohang Iron and Steel Company (which would be renamed POSCO).

Park Tae-joon was born in 1927 in Kyungsang Province in Korea and moved to Japan when six years old, where his father was working. In 1945, he enrolled in Waseda

University's College of Science and Engineering. But the war ended in 1946, and Park Tae-joon returned to Korea joined the military, graduating in 1948 from the Korea Military Academy. When the Korea War began in 1950, Park Tae-joon was promoted to the rank of Major. In 1953 he was promoted to Chief of Staff of the Fifth Division of the Korean Army: "Although Park demonstrated many talents as he moved up the ladder in the military organization, he was recognized and rewarded for one particular skill – an uncanny ability to manage and motivate people. On February 7, 1960, he was appointed Staff Officer of Army Personnel Affairs." (Innace and Dress, 1992, p. 25)

Park Tae-joon did not participate in General Park Chung-hee's military coup on May 16, 1961. But their relationship remained close: " . . . there was a very close and special bond between Park Tae-joon and Park Chung-hee. . . . The relationship began . . . when they were in the Military Academy. Park Tae-joon was a cadet and Park Chung-Hee was a company commander. . . the elder Park took a special interest in his prized pupil. The commander singled out Park Tae-joon for unusual duties even after he had graduated in July 1948. And in 1961 when General Park became the nation's leader, he counted on Park Tae-joon for special assignments. 'There are many things that I learned from him', said President Park, 'particularly his knowledge of economics.'" (Innace and Dress, 1992, p. 26)

In 1963, Park Tae-joon retired from the military; and in 1964, President Park sent Park Tae-joon to Japan to observe the Japanese steel industry. On his return, Park Tae-joon was appointed president of the Korea Tungsten Mining Company, then on the verge of bankruptcy. He turned the company around to profitability.

In November 1967, President Park established an government committee for constructing the steel company and appointed Park Tae-joon as Chairman of the committee. The committee was to formulate a business plan and to have construction blueprints designed for an integrated steel company. President Park Chung-hee then appointed Chairman Park Tae-joon as manager of the new steel company, government sponsored but privately run. Park Tae-joon decided to accept the management appointment but only under conditions: "Chairman Park Tae-joon jotted down three conditions he believed were absolutely vital for the steel mill's success. If these terms were granted, he would accept the job. . . Nervous, but firm in his convictions and armed with his note, Chairman Park made his way to the Blue House (home of the president of South Korea). It was winter in late 1967, but the chill running down his spine was not from the cold alone. His pending meeting with the President filled him with apprehension. . . he entered the room where Preident Park waited for him. He presented his three conditions. . . . the President read them over slowly, . . . President Park looked up at Park Tae-joon and smiled . . . he jotted his initials. The deal was done. The project would proceed according to Chairman Park's terms." (Innace and Dress, 1992, p. 40)

These were the three conditions. First, that POSCO as a private company (although initially financed by government) would have absolute control over procurement of equipment, goods and services. No government agency would procure anything for the company. Second, POSCO would have absolute control over staffing. No government agencies would assign personnel to the company. Third, POSCO would not make any donations to political parties for their campaigns. Political parties would not be allowed to 'squeeze' the company for campaign expenses. Krauk Jeung, one of the other founders of POSCO, wrote: "'This was a contributing criterion to his (Park Tae-Joon) subsequent success with the founding of POSCO. He demanded a high level of integrity and honesty from his people. He did not brook outside interference or tolerate 'backdoor' promotions . . . And by the time POSCO was founded, he got those assurances in writing from National President Park for a corruption-free system for POSCO so that there would be no outside influences." (Innace and Dress, 1992, p. 41) Integrity, honesty, corruption-free system – such were the managerial values upon which Park Tae-Joon would build POSCO.

So far in this case, we can observe three management-of-technology criteria concerning the entry of Korea into the international commodity business of steel. *First, the needs of the country and a government commitment to establish the steel industry were both vital to the industrial development of the nation. Second, there was need for the government to provide initial financing for the construction of the manufacturing plant (because of the enormous investment required to enter the international steel industry in the 1960s). Third, the company (although government sponsored) would operate from the start as a private company – managed as corruption-free, with integrity and honesty.*

Yet later in 1968, the international community changed their collective mind about helping to finance Korea's steel company: "But in November, a report issued by the World Bank . . . urged the Korean government to give its irs priority to the machinery sector, instead of steel. Korea had no knowledge of integrated steelmaking. Since Korea was not in a stable enough position to repay loans in the long-term, the machinery industry was a much safer bet for the economy, and, of course, for its potential financial backers." (Innace and Dress, 1992, p. 58) Following this report a final meeting was held in February 1969 in Pittsburgh by the international group. They told Chairman Park the group had changed its mind and would not help finance the project: ". . . a Koppers executive, notified them (the Korean delegation) that the West German and British governments were doubtful about the need for, the priority of, and the feasibility of the steel mill project . . . ." (Innace and Dress, 1992, p. 59) .

Devastated by the turn-around, Chairman Park, headed home. But before returning directly to Korea with the bad news, he decided to pause for a couple of days in Hawaii – and think. There he had an idea. "He knew that Japan was obligated to pay reparations compensation to Korea because of its colonization of the country for 35 years. . . . Park wondered how he could divert some of that Japanese compensation to the steel mill project." (Innace and Dress, 1992, p. 40)

Perhaps this could be a solution to finance the project! Chairman Park returned to Korea and presented his idea to President Park – to divert some of the reparation monies to the project. President Park agreed with the proposal, provided the Japanese government would agree. Chairman Park flew to Japan to try to gain their approval: "Upon arriving in Tokyo, Park Tae-Joon called upon the Confucian scholar Seitoku Yasuoka, as well as the President of the Japan-Korea Cultural Association, Nobuo Yagi. In particular Yagi, had considerable power within Japanese government and business circles. . . . Both Yasuka and Yagi agreed to help Park. They offered to introduce him to the President of the Japanese Iron and Steel Federation Yasuhiro Inayama, who was also President of Yawata Steel. Park gladly accepted, and a meeting was quickly arranged." (Innace and Dress, 1992, p. 40)

In Korea, President Park persuaded the legislature to approve of a partial use of the reparation funds for steel production. President Park's government then guaranteed the financial backing of loans for the steel project. The Japanese government agreed : "The total foreign capital required was pegged at \$123.7 million; of this amount \$30.8 million would come from the reparations monies;, \$42.9 million form a Japanese OECF loan, and Japan's Ex-Im Bank would finance the \$50 million balance via a deferred low-interest loan." (Innace and Dress, 1992, p. 40)

We can see from this case that capital is a significant barrier to entry into an international commodity industry at a competitive level. The Korean government could not afford to finance the new Korean steel industry without foreign loans. This is an example of the general principle that foreign capital has always been and continues to be critical to national industrial growth. In the case of POSCO, the European and American steel companies did not want to see another major steel competitor enter the international market. In contrast, the Japanese government agreed to help finance the Korean steel project to improve the international relationship between Japan and Korea (after that last half-century of occupation of Korea).

As is well known, the management job of a leader is (1) to acquire technology, (2) to formulate an executable plan, (3) to acquire capital, and (4) to assemble a team of personnel. With the financing accomplished, the project could begin. Park had arranged to acquire technology and equipment from Japanese equipment manufacturers and steel makers. He had engineering plans designed and checked for the construction of a large integrated steel plant. Now he had arranged for financing. The next tasks were to assemble and train personnel, prepare a site, and construct the plant.

Organizations are led by a chief executive (president), but personnel under the leader do the actual work. The personnel must be capable of doing the work and so must be trained. Park assembled 40 initial members of his team and sent them for training: "He told each of the members leaving for training in Japan that they should leave no stones unturned. He instructed the overseas training teams to acquire every bit and piece of the host nation's technology. . . . Upon their return, the trainee-delegates were immediately put in charge of internal training programs." (Innace and Dress, 1992, p. 40)

The site of Pohang was chosen because it could have good harbor facilities. The first part of the construction was to improve the harbor for large ships, eventually as large as 250,000-ton capacity ships – to bring in iron ore and take out steel products. The second part was to level land for the construction of the huge steel works.

The construction of the steel-making capacity was planned in stages, with the first stage to produce 600,000 tons of steel products annually: "POSCO'S plan was to introduce steel product rolling first, followed by steelmaking, and blast furnaces. . . . (With rolling first, this would give POSCO) the option of purchasing semi-finished steel slabs and rolling them into marketable end-products ... (giving) POSCO an immediate cash flow with an opportunity to achieve profitability faster." (Innace and Dress, 1992, p. 40).

Chairman Park planned to achieve sales, revenue and profitability quickly. This emphasis on 'speed': speed-of-construction and speed-to-market was the distinguishing characteristics of Park's management style. In addition to technology-plan-capital-personnel, one can see that also the ideas of 'resources' and 'market' were essential criteria in Park's leadership.

POSCO had to import iron ore from Australia: "Australians asked for a higher price per ton than the Japanese were paying . . . When questioned about the price they argued that Japan was already producing 100 million tons of steel, while Korea was planning only one million tons. . . . Park did not like that at all . . (it) would put POSCO in an unfair competitive position. . . He decided not to sign a contract . . . and began to explore other material supply options, in the U.S., India, and some other countries. . . The Australian raw material suppliers were getting anxious, . . And soon afterwards, they agreed to supply POSCO at the same price per ton . . ." (Innace and Dress, 1992, p. 97)

As a private company, Chairman Park kept a focus on controlling costs to be competitive; and we can add to Park's management thinking of technology-plan-capital-personnel-resources-market, the ideas of 'speed' and 'revenue' and 'competition' and 'profits' and 'cost-control'. The demand in Korea for steel was large enough for POSCO to profitably sell products against competitors when priced competitively. The keys to Park's leadership in starting POSCO now add up to*: integrity-technology-plan-capital-personnel-resources-market-speed-revenue-competiton-profit-costs.*

Construction began on the first facility of the integrated steel plant: "The foundation work for the hot strip mill was started in April 1971, but sometime in July, it began falling behind schedule . . . (By August) construction was running more than three months late. . . . That was when Park stepped in. He returned immediately to Korea, and by August 18, he was in Pohang inspecting the site . . On August 20, the Chairman declared a state of emergency. He mobilized all the construction squads to focus their efforts on the hot strip mill foundation. . . . All managers at the level of assistant director were ordered to take turns serving as the chief supervisor of the day. 'We were required to lay 700 cubic meters of concrete each day, come hell or high water . . .'" (Innace and Dress, 1992, p. 105)

Steel equipment is gigantic and very heavy and so requires enormously large foundations upon which to rest. Without a good foundation, the steel production equipment is not secure. Chairman Park insisted on sticking to schedule. When the schedule was not being met, Park declared a state of emergency and had all POSCO executives down on the construction site personally supervising work. Everyone was involved to fix the situation. We can add to Park's management style a focus on keeping to the schedule and making executives personally deliver. So leadership style criteria in the case of POSCO include*: integrity-technology-plan-capital-personnel-resources-market-speed-revenue-competiton-profit-costs-schedule-supervision.*

Another problem occurred: "With the foundation poured, erection of the steel frame for the mill's steelmaking plant was begun in the summer of 1972. . . One morning, on one of his regular inspection tours, Park decided to go to the top of the steel-framed structure . . . From that perch, . . . he spotted that many of the heads of the bolts remained on the steel frame. This meant the fastening of the structural steel was loose in some spots and, as such, the building would not be able to support the equipment and machinery inside the mill . . .Park called in a team of his white-collar managers to assist with the trouble-shooting. . . There were hundreds (of bolts) that need to be fixed. Had it not been for the Chairman's sharp eyes, a serious accident could have stricken the mill shortly after it was up and running." (Innace and Dress, 1992, p. 106). To Park's leadership style criteria, one can add problem-detection-and-solving, or 'solutions': *integrity-technology-plan-capital-personnel-resources-market-speed-revenue-competiton-profit-costs-schedule-supervision -solutions.*

The first stage of Park's construction of plants was completed on July 4, 1972. The plant could finish steel slabs into sellable products; and POSCO purchased semi-finished slabs from other steel producers to make steel plate. It sold that and posted a profit in its first year of operations. The second stage of Park's plan was to begin the construction of steel making furnaces (making steel from iron ore). But additional financing for this stage was needed, and further financing solely by the Japanese was not likely: ". . . Chairman Park wanted to demonstrate that POSCO could expand without the Japanese, and that the Europeans . . . and many others wanted a piece of the action. . . (then) Japanese companies (wanted to) be included in the negotiations for second stage . . . . Proceeding in this fashion, many of the facilities for the second stage were obtained at rock bottom prices." (Innace and Dress, 1992, p. 116).

Park used the latest of technologies. Computerized process control of production of steel was installed. Construction of the blast furnance plant was completed on November 30, 1978: "A steelworks operates around the clock, 24 hours per day. Churning out its molten metal on one end and solid (steel) products on the other, in a constant process flow. . . Raw material conveyors rumble from the dockside yards, making their way to the steep incline . . to the tops of the blast furnaces . . . On the furnace decks, wokers watch the steady flow of red-hot pig iron." (Innace and Dress, 1992, p. 116)

A third and fourth stage of expansion of production capacity occurred through 1979: "The late 1970s were boom years for the Korean steel industry. POSCO was making money hand over fist." (Innace and Dress, 1992, p. 116)

On 26 October 1979, President Park Chung-Hee was assassinated by his head of the Korean Central Intelligence Agency. General Chun Doo-Hwan arrested and executed the assassins; and he assumed the presidency. However, Chairman Park Tae-Joon continued in the management of POSCO and expanded its capacity.

In 1991, POSCO was the third largest steel company in the world, with a capacity of producing 17.5 million tons of steel annually.

The combination of good management, advanced technology, national need, and corporate culture enabled Chairman Park to succeed: "The strength of POSCO's corporate culture is rooted in a commitment early on to its concept of one family – a community spirit . . . POSCO has demonstrated a caring concern for the well-being of all its people since the very beginning." (Innace and Dress, 1992, p. 116) To Park's leadership style criteria, one can add 'culture': *integrity-technology-plan-capital-personnel-resources-market-speed-revenue-competiton-profit-costs-schedule-supervision –solutions-culture.*

PRIME-MOVER

We are calling a successful late-entry into a commodity market a 'prime mover' (in analogy to Chandler's label of a successful first-mover in a new-technology industry). Using this case of POSCO as an exemplar of a prime mover, we summarize the criteria for a prime mover management style, Figure 3.2.

Six major criteria need to be focused for a prime mover to enter an industry with commodity-type products: Technology, Capital, Market, Competition, Plan, Integrity. All are standard criteria for the management function in business literature, so there is nothing new about each of these. But what is new is how they must interact together for a successful prime move into a major commodity industry.

In the criterion of Technology, an industrial competitive-entry must use the most advanced technologies available to the industry – as the products of a prime mover must perform equally well as any products in the industry and with highest quality. Second-rate quality will not buy competitive entry. Performance and quality establish brand reputation of the new competitor. But to acquire and install the most-advanced technology, many technical problems will occur and must be solved by the personnel. While advanced technology equipment can be purchased, knowledge of how to use the technology for high quality cannot be bought. It must be learned quickly by the personnel of the prime-moving company. Also resources to be processed to make a product must be the accessible to the new company, otherwise products cannot be made in quantity. *For a prime mover, technology as advanced as any established competitor must be acquired and learned and implemented by the newly entering firm, along with access to resources.*

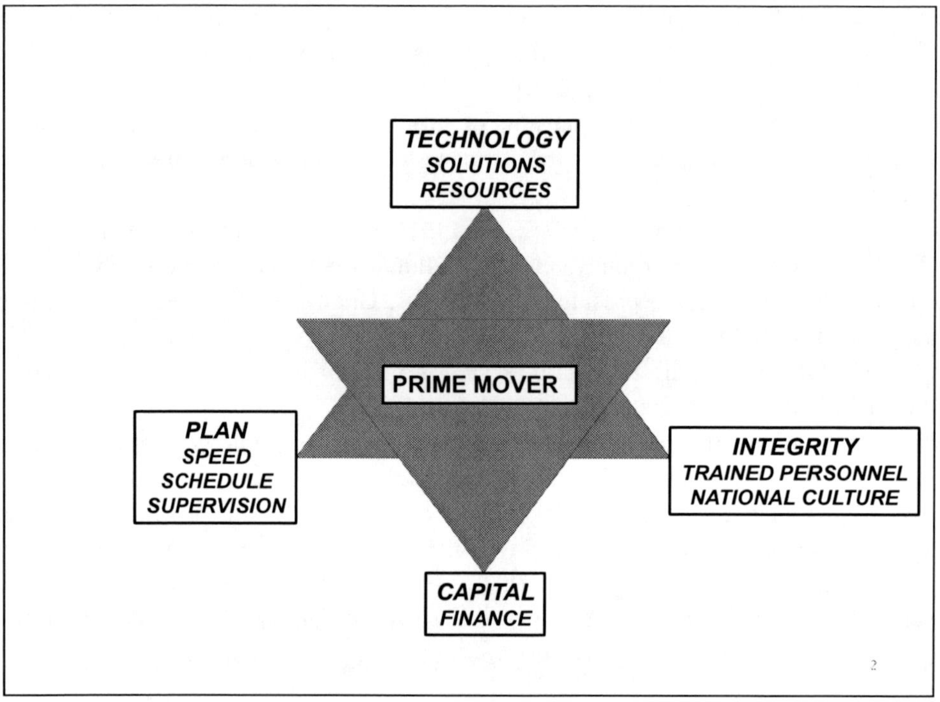

Figure 3.2. Prime Mover in a Commodity Industry.

In the criterion of Capital, a very large investment is required for competitive entry, because established firms in the industry already have large production capacities and major marketing capabilities. Investment in production and marketing are always the major investments in hard goods. But even in services, investments can also be large in facilities, delivery capacity, marketing. Large investment to enter the markets of an established industry is the basis of what economists have called 'barriers to entry' (see for example, (MacAfee et al., 2003)). In financing a new entry into a major competitive commodity product, it is difficult to obtain sufficient capital. As we saw in POSCO, without the South Korean government's strategy and commitment, financing for a new major integrated steel maker in Korea would not have been possible. The international banking and steel industries thought another international steel competitor unnecessary for the world (even if it were desirable for Korea). *For a prime mover, large initial capital is needed to finance construction of production facilities and make initial sales; successful entry into a commodity-type market is expensive.*

In the criterion of Market, national markets will be large for commodity-type products of a technology-mature industry. For a new entry to a commodity market, it is difficult to take market-share from established competitors because they have the advantages of brand recognition and established distribution and retail capabilities. In the case of POSCO, market entry was easier because the first market for new steel producer was the growing market of South Korea. The demand for steel in South Korea made it easy for POSCO to initially sell all its steel products, making revenue and profits from the start. Only later, after POSCO had increased production capacity to a very large scale did it need to compete in the international market for steel. *For a prime mover, markets need to be growing, so that a new entry can initially capture market share without having to first displace established competitors*

In the criterion of Competition, possible profits for a new entry are bounded by the prices of the existing products of competitors. Entries should not try to enter with higher prices but with lower prices. With established product prices as an upper price barrier, profits are determined by costs. Initial low costs are necessary for profitability. Thus a focus upon low costs must be accomplished first and early, because having the time to learning how to lower costs (called the 'production-learning-cove') is not a luxury available to a new entrant. For a prime mover, cost targets have to be built into an initial business model and met. If competitive costs cannot be obtained initially in constructing the production facility and in designing the product, the new entry will not survive against established competition. *For a prime mover, initial costs of production must match or be lower than the production costs of established competitors – and at good or higher quality.*

In the criterion of Plan, a prime-mover must have three elements of its initial business plan: (1) product development and design, (2) construction of production plant, (3) customers and initial sales, (4) financial projections of operations when production begins. In the case of POSCO, its first planned products were steel plates rolled from steel slabs and sold in the Korean market. It's plans for construction of the facilities was in four stages: first the rolling mill for producing steel products, then the blast furnace for producing slab steel, then expansion of facilities for increased production (in stages three and four). Because of the large but finite amount of financed capital, it was essential to keep the construction of facilities on time and on budget in order to make it through the first stage and second stage to an integrated steel plant. The customers for the first years were Korean industries and, for later years, international markets. Speed was important to keep to schedule. For this, finding and solving problems quickly was necessary to finish the project on time. Additional financing would not have been available to POSCO, until after POSCO successfully finished stages one and two. Then it could hope to raise additional financing for expansion of capacity in stages three and four. Close supervision of all operations directly by senior level managers, and while on-the-ground, was essential to quickly find and fix problems. Problems always occur in any project and can delay schedule. . *For a prime mover, good plans and quick implementation is necessary to enter a commodity-type industry, competitive right from the start in quality and cost.*

In the criterion of Integrity, four levels in the organization needed to act honestly, competently, and with responsible transparency: government, executive, management, and worker levels. At the governmental level, government regulation of an industry is proper for safety and financial reporting. And in the case of POSCO, government planning for Korean export-industry was necessary to create the initial financial investment for the construction of POSCO facilities. It had been important that President Park Chung-Hee agreed to keep government from interfering with operation of POSCO as a private business. Chairman Park Tae-Joon had the integrity to run an honest, effective business with proper care and support of its workers. At the managerial level, POSCO managers respected integrity and managed with competent supervision, ensuring that schedules and targets were met. They were required to say in touch with the day-to-day business of the work, especially during critical times. At the worker level, employees worked hard and long hours, receiving decent salaries and believing in the integrity of POSCO's mission to help build the South Korean nation. *For a prime mover, integrity, honesty, competence, responsibility, hard work are necessary in all employees for a new firm, to compete immediately with established competitors.*

In summary, good management of any enterprise should have advanced technology, sufficient capital, growing markets, high-quality-low-cost production, good plans and implementation, and complete integrity. But what is different in case of a prime-mover is that all these six qualities must be embedded immediately in the organization – with no time to learn them eventually. Therefore as in any entrepreneurial event, initial leadership is vital and critical – giving life and shape to the new firm. There is no luxury of learning how to lead. The character of a leader is the life or death, initially of the enterprise.

For example, Joseph Innance and Abby Dress emphasized: "Success was paramount to Chairman Park, as well to President Park. And Chairman Park did not want anything to stand in the way. He needed to eliminate all obstacles that might present a problem to the project. The major obstacles, in Chairman Park's mind stemmed from his jaded, but accurate view, of government practices and improprieties – not just in Korea. In many developing countries at that time, government corruption was rampant. Corruption doomed many key project and hampered national progress. . . (So) Chairman Park . ... requested a face-to-face meeting with the President." (Innace and Dress, 1992, p. 39) This was when Chairman Park asked President Park to sign off on a agreement of government noninterference in POSCO – no interference in procurement (a major point of corruption), none in personnel appointment (a major source of incompetence), and no required political donations (a major form of corruption). Thus leadership strategy, commitment, competence, and integrity were all critical personal elements in the two Parks, President and Chairman. Their characters of leadership were essential to the success of POSCO. Figure 3.3.

Figure 3.3. POSCO.

Industrial Life-Cycle of the Steel Industry

We can now construct the industrial life-cycle of the steel industry to examine the industrial context in which POSCO was begun in 1968. The modern steel industry began in 1854 with the innovation a new process for producing steel, invented by Henry Bessemer. The key point of the invention was that it used coke (partially burned coal) to melt the iron ore, instead of charcoal (partially burned wood). Coal was abundant and cheap, and, upon its use as coke for steel-making, the industrial revolution began (along with innovations in the steam engine and textile machinery). Partially burning the coal burned off impurities (as did partially burning wood) and did not introduce them into the molten iron. Also Bessemer blew air into the molten iron to burn off impurities in the iron ore and reduce excess carbon in the ore. It was chemical (impurities) in the coal and iron ore that had to be removed to make good quality iron; and after which a proper amount of carbon was added into the molten iron to make good quality steel. The steel industry began not upon a product innovation but upon a process innovation. As a product, steel had been invented back in beginning of the Iron Age of iron, around 1500 BCE. Thus the life-cycle of the modern steel industry began in 1850 with production innovations. More major production innovations followed: in 1875 with the Siemens-Martin process (open-hearth furnace instead of Bessemer's refractory barrel) and in 1950 with the Linz-Donawitz process of blowing oxygen into the molten iron instead of air. A reason that POSCO could successfully enter the steel industry in the 1970s was that the basic core technologies of product and production were well known and outside patent lifetimes. Figure 3.4.

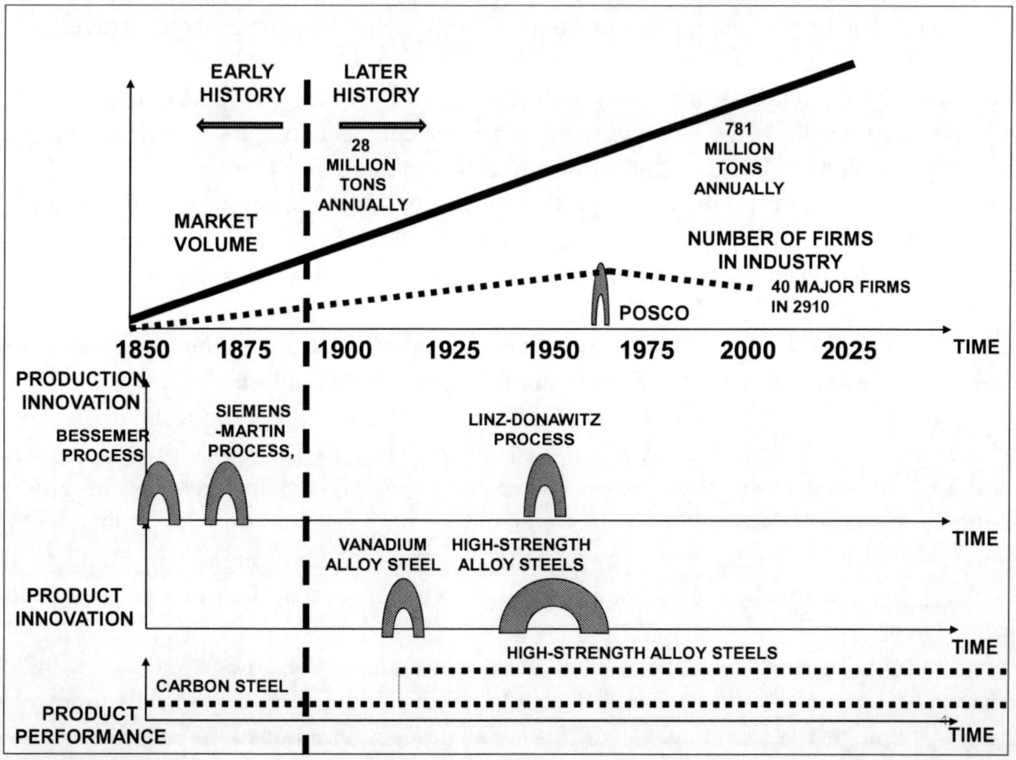

Figure 3.4. Life-Cycle of the Steel Industry.

Also the market volume of the steel industry never leveled but continued to grow, even as a mature-technology commodity industry. Steel is an essential material to industrialization and economic development around the world. In 1900, world steel production was 28 million annually; and a hundred years later in 2000, production was 781 million tons annually, and still growing. POSCO could successfully enter the steel industry because the world's steel market was still growing. Also the local Korean market for steel was exploding, as South Korean began industrializing after 1963.

In terms of product innovations, high-strength alloy steels were innovated in the 1920s and afterwards – but long before POSCO was established. Also in terms of resources, iron ore and coal, there were still in the world sources of iron ore and coal not fully utilized by the nation mining them. For example, Australia was a major source of iron ore in Asia when POSCO began.

In summary, technology was not a barrier to POSCO's late entry. The barriers were capital and management and resources. The economic idea of economy-of-scale was a barrier to entry, but only in capital and resources. (At a smaller scale, a steel industry of mini-steel mills began in the late 1960s for recycling steel from scrap steel.)

HIGH-TECH AND COMMODITY INDUSTRIES

High-tech products are products that are rapidly improving in technical performance due to continuing technical progress in the core technologies of an industry. But as the core technologies in an industry mature, products become relatively undifferentiated in technical performance. Then price and quality of the product become the primary competitive factors. In contrast to high-tech products, performance-undifferentiated products are called 'commodity' type products – since they all technically alike in terms of performance.

Industries are called *high-tech industries* when the core technologies in their products and production are rapidly changing. Industries are called mature, or *commodity industries*, when the technologies in their products and production are no longer changing rapidly or much at all. Businesses in high-tech industries can compete upon superior technical performance of product, whereas businesses in commodity industries can only compete upon price and quality.

In the previous Figure 3.1, one can see the direct relationship between technology finiteness and industrial maturity. Therein are the charts for (1) industrial life-cycle, (2) rates of product and process innovations, and (3) technology S-curves for core technologies of the product-system. Looking along the time dimension, we see that the major growth in a market in the applications growth phase begins when the design is standardized for the product. Eventually thereafter, the industry begins a mature technology phase for its product and the product becomes a commodity-type product as opposed to a high-tech product. The technology S-curve for the core-technology to the product levels off as product innovations stop.

Yet during this time, technology innovation continues in the production process, and a commodity-type product still uses high-technology, technological change, for its production. During this time, market-volume can still grow, as production innovations reduce the cost of producing the product. As the process innovations decline, then the technology S-curve for the production processes levels off. Only after both product and production innovations have declined, does one have a totally technology-mature industry. Then until a substituting technology occurs for the core-technology of the product or for the core-technologies of production, the technology-mature industry can continue at a market level determined by market replacement rates and/or market demographics. Over time, all core technologies mature; and all high-tech industries eventually become commodity industries.

The profit margin for any firm's product lines are determined by two directions of conditions, internal and external. The internal conditions consist of the efficiency of the firm and the strategy of management. The external conditions are the balance of supply to demand in the industry.

The external factors include the number of competitors, how rapidly technology is changing, and how rapidly the market is growing. When an industry reaches technological maturity and products are commodity-type, then the profitability of any firm in the industry is bounded by the external conditions of the number of competitors and production capacity of the industry. High-tech products can provide high gross margins because of the limited competition in the technical performance high-tech products provide. But, the profit margins of commodity products are set by industrial competition only on price.

There are two basic business strategies concerning technological innovation: first a high-tech-product strategy and later a commodity product strategy, the high-tech-product strategy is time-limited. At first all businesses in a new industry are high-tech businesses, but over time all businesses in that industry become commodity businesses. *As an industry matures from high-tech toward a commodity industry, the global competitors which survive are those companies: (1) who are low-cost, high-quality producers, and (2) have established national and international distribution capability.*

When a new core-technology is invented and substitutes in a commodity industry, then the previous producers in the industry become technologically obsolete. Any industrial sector that then continues producing a technically obsolete product-line will become industrially obsolete. The societal function of an industry never becomes obsolete, but the core technologies of an industry can become obsolete for that function. An industry becomes obsolete and dies when a substituting technology occurs to replace the core technologies, and the industry does not change to the new substituting technologies.

LESSONS LEARNED

We have seen in the history of POSCO the success of a type of leadership in commodity industries in a global context – the style of leadership of a 'prime-mover'. This is a management style essential for a successful competitive new entry into an international mature-industry market. This is important to economic development, because developing countries have more often entered mature markets than begun new high-tech markets. The following criteria in management style as a 'prime-mover' are important:

At first all businesses in a new industry are high-tech businesses, but over time, all businesses in that industry become commodity businesses.

As an industry matures from high-tech toward a commodity industry, the surviving global competitors are those: (1) who are low-cost, high-quality producers, and (2) have established national and international distribution capability.

In entering global competition in a commodity industry, technology should be as advanced as that of any established competitor, both learned and implemented quickly by the newly entering firm.

Large initial capital is needed to finance construction of production facilities and to make initial sales, as successful entry into a commodity-type market is expensive.

Markets need to be growing in a commodity-type industry, so that a new entry can initially gain market share without having to first displace established competitors.

Initial costs of production of a new entrant must match or be lower than the production costs of established competitors.

Good plans and quick implementation are both necessary to enter a commodity-type industry, being initially competitive in quality and cost.

Management integrity (honesty, competence, responsibility, hard work) is necessary in all levels of employees, for the new firm to compete immediately with established competitors.

Park Tae Joon
Source: (http://www.amm.com/HOF-Profile/ParkTaejoon.html)

Chapter 4

LEADERSHIP IN A GLOBAL CIVILIZATION – KOREA

INTRODUCTION

As we have seen in the cases of Samsung and Posco, the relationship of industry to government in South Korea has been important, an essential factor in the growth and competitiveness of the Korean economy. Overall, this relation was nicely summarized by Joonghae Suh and Derek H. C. Chen: "The role of the government in the Korean economy is not to substitute for, but to complement, the functioning of the market." (Suh and Chen, 2007). And this complementary relationship evolved. We will review the history of the government of South Korea in the period from 1945 to 2013, in order to understand its relationship to the industrial development of Korea. This period can be analyzed in a cross-disciplinary framework of societal dynamics (Betz 2010) as a series of events which alter the state of South Korean society, shown in Figure 4.1.

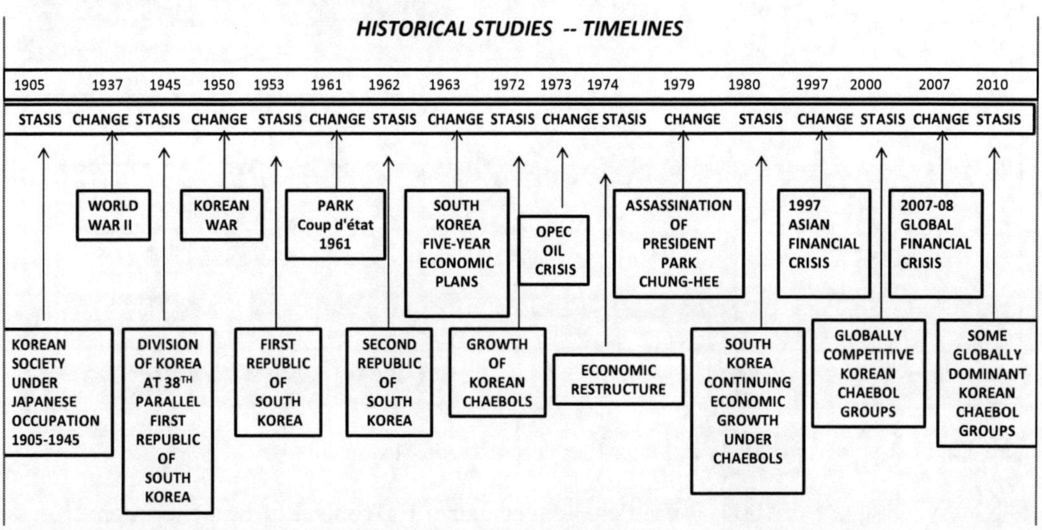

Figure 4.1. Change Events in Republic of South Korea.

HISTORY: 1894 TO 1945 KOREA

The history of modern Korea begins, unfortunately, with its occupation in 1905 by the then new Empire of Japan. How had Japan created its empire, of which Korea was the first conquest? A key historical event leading to the occupation of Korea was a Sino-Japanese War in 1894-95, in which the new government in Japan (called the Meiji restoration government) fought the Chinese government (under the last Empress of the Qing Dynasty) for control over Korea. In six months, the modernizing Japanese army and navy won battles over the Chinese military. This was due to Japan's success in reforming its military to European standards and China's failure to do so. The battles are depicted in Figure 4.2. For control over Korea, the first battle occurred in July 1894, when 4000 Japanese soldiers landed at Incheon, near Seoul, and attacked 3500 Chinese soldiers, garrisoned near Asan. The Chinese garrison fled north to Pyongyang, joining additional troops from China. In September, Japanese forces marched north to Pyongyang, circled the city and attacked. The Chinese soldiers surrendered and, when let out of the city, fled north. After severe losses, on land and at Sea, the Chinese government sued for peace. Some Japanese forces stayed in Korea, influencing the Korean government but not yet completely controlling it.

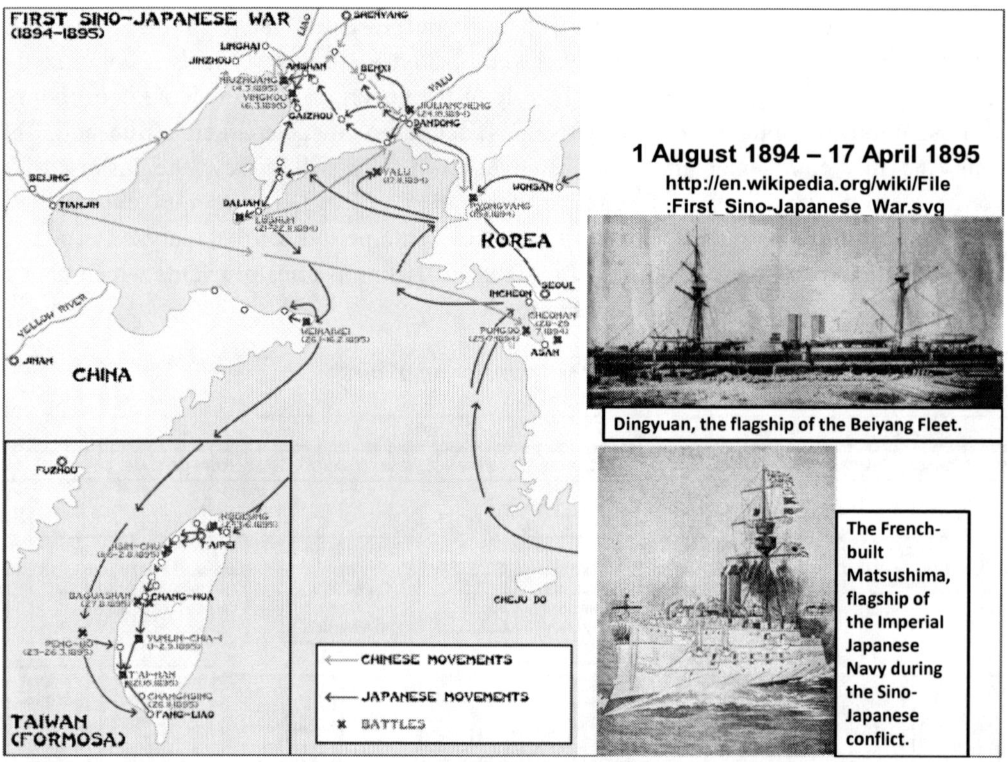

Figure 4.2. First Sino–Japanese War, Major Battles and Troop Movements.

Nine years later, in 1904, Russia challenged Japan for control of Manchuria and Korea, the 'Russo-Japanese War'. Russian had constructed a port at Dalian, which the Russians called 'Port Arthur'. It would provide the Russian empire a year-round ice-free harbor for its Siberian territory. The Japanese interest in taking Dalian was to acquire for Japan the beginning of an Empire on the Asian continent. In February 1904, war began between the Russian Empire and the Empire of Japan.

On the 8th of February 1904, Japan issued a declaration of war to the Russian government and attacked Port Arthur the same day. The Japanese fleet attacked the Russian battleships, the Tsesarevich and the Retvizan, with a new weapon, torpedoes. The Japanese Navy damaged the ships, but was unable to conquer the port. For this, the Japanese Army needed to attack from the landward side, from the north of the port. Next the Japanese Army landed in Korea and went north and around to Dalian and attacked from the landward side. All the time it took for the Japanese Army to reach Dalian, the Russian ships had been bottled up in the harbor by the Japanese fleet outside the harbor. The Russian Tsar had decided to send the Russian fleet (in the Baltic Ocean) around the world to China to relieve the ships bottled up in Port Arthur. It took from May to August to sail around Africa and across the Indian Ocean into the South China Sea and up into the Tsushima Straights between Korea and Japan, Figure 4.3.

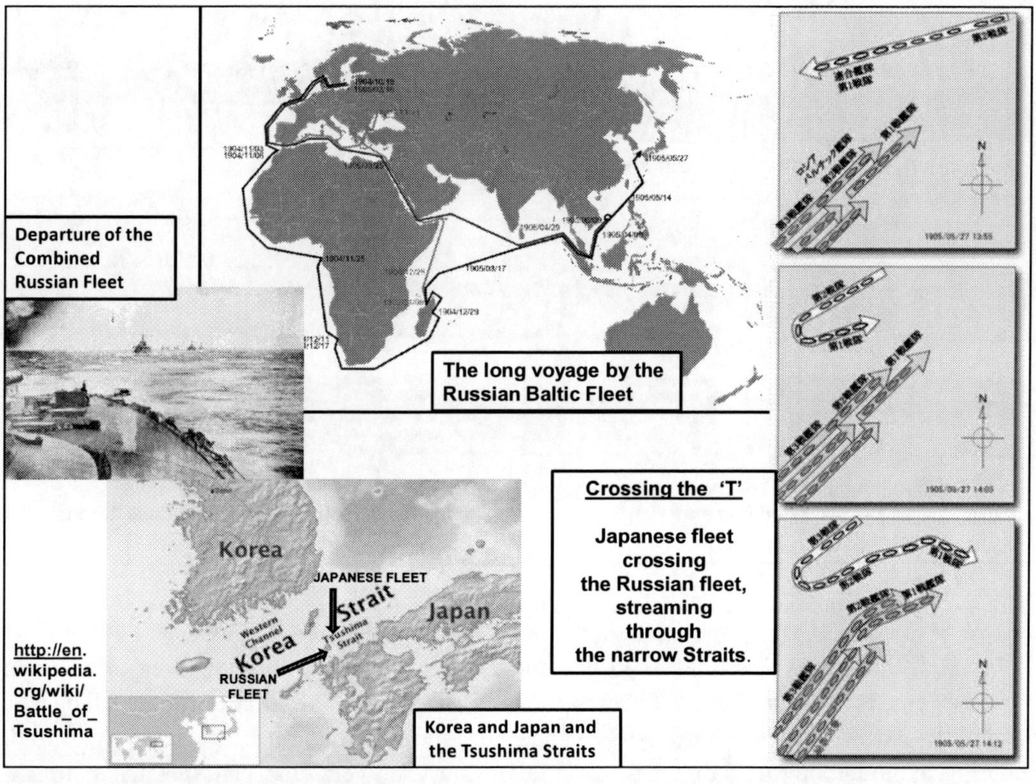

Figure 4.3. Defeat of the Russian Fleet 1905.

All the time, Japanese military intelligence followed the progress of the Russian fleet, telegraphing information on its progress to the Japanese Navy. Thus the exact path and date when the Russian fleet would enter the narrow waters of the Tsushima Straits were known by the Japanese Navy. It had set a trap for the Russians. As the Russian ships had to sail one after the other through the narrow straight, Japanese battle ships sailed across the outlet of the straights, aiming their guns on each emerging Russian. Since battleships then had more guns aimed sideways from their broadside then guns aiming forward, the Japanese battleships in that battle had much more firepower focused upon the Russian battleships, than had the Russians ships on the Japanese. The Japanese battleships effectively 'outgunned' the Russian battleships in the maneuver at Tsushima Straits. This maneuver is a classic naval battle formation, called 'crossing the T', shown in Figure 4.3. The Japanese battleships sailed across the path of Russian battleships locked in by the straits. On May 27-28, 1905, the Japanese Admiral Togo Heihchiro destroyed two-thirds of the Russian Fleet under Admiral Zinovy Rozhesvensky. Russia had lost the Russo-Japanese war.

The importance of technology in the changing history of the world is illustrated by the progress in battleship technology to 1905, Figure 4.4.

Le Napoléon (1850), the first steam battleship

The French La Gloire (1859), the first ocean–going ironclad warship

The French Redoutable (1876), the first battleship to use steel as the main building material

Imperial Japanese Navy's Satsuma 1906 , designed and built by Yokosuka Naval Arsenal

http://en.wikipedia.org/wiki/Battleships

Figure 4.4. Progress in Battleship Technology 1850 – 1906.

In 1905, the Japanese military had become a modern world military power, after having in 1868 begun to wholly change Japanese society, under the 'Meiji restoration'. Japan had changed its social structure and governance to incorporate modern technology, rapidly becoming a modern industrialized nation and military power. One can see that battleship technology had been rapidly changing from 1850 to 1906, with the addition of steam propulsion technology, ironclad construction, steel construction, and large steel guns.

In 1905, Russian and Japanese military battleship technology was equal; so that when Russia and Japan clashed in 1904-05, both governments had equal military technology. The difference in the conflict was about leadership, strategy and military competence – not about weaponry. The leadership of the Tsar of Russian conceived of poor strategy in sending the Baltic Fleet around the world; and the leadership of Admiral Zinovy Rozhesvensky was incompetent in tactics – guiding his fleet through narrow straights, right off the coast of

Japan. Instead of being trapped in the Tsushima Straits, Rozhesvensky could have sailed straight to Dalian to relieve the trapped Russian fleet in that port.

When societies are equivalent in technology, then good leadership in one society triumphs over poor leadership in another society.

Thus in 1905, Japanese military had occupied Korea. Japan expelled Chinese influence over Korea and then expelled Russian influence over Manchuria. In 1910, the Japanese government formally annexed Korea into the new Japanese Empire.

Japanese settlers had been entering Korea since the First Sino-Japanese War in 1894 and by 1910 had reached 170,000 settlers. The Japanese government encouraged Japanese farmers to emigrate and Japanese businessmen to establish businesses in Korea. In 1910, Japanese ownership of Korean agricultural land was around 7-8% but grew to about 53% by 1932. The Japanese government introduced a public education system (which had been introduced earlier in Meiji Japan) In the 1940s, 32 % of Korean children were obtaining an elementary education. The curriculum taught the history of the Japanese Empire and inculcated reverence for the Emperor, and the language of instruction was Japanese.

The next critical event in the history of the Japanese occupation of Korea was the conquest of Manchuria. Manchuria had been under the control of the Qing Dynasty of China, since its founding in 1644. The Chinese Emperors of the Qing were Manchurian in origin. In 1858, the Russian expansion into Siberia, enabled the Russian Tsar to control the northern part of Manchuria. In the 1890s, Russian influence extended over the southern Manchuria and built the Chinese Eastern Railway from Harbin to Vladivostok, Figure 4.5.

Figure 4.5. Map of Chinese Eastern Railway (Trans-Manchurian).

After the defeat of the Russian fleet in 1905, Japanese influence replaced Russian control over southern Manchuria. After the abdication of the Russian Tsar in 1917 and the Russian Revolution, Japan moved into northern Manchuria. Later in 1925, after new Russian forces under the communist government attacked once again in Manchuria, the Russian Soviet government regained control over northern Manchuria.

Six years later in 1931, a Japanese Kwantung Army, invaded northern Manchuria and formally established the puppet state of Manchukuo. Japanese immigration into Manchuria increased from about 100,000 immigrants in 1932 to about 850,000 in 1945. Economic growth occurred as Japanese firms launched new enterprises. Raw materials were exported to Japan. The Japanese army cooperated with Japanese industrialists in build an industrial base for military weaponry, in just the same societal model which had modernized Japan after the Mejii restoration. In 1939, Manchukuo and Korea and parts of China were central territories in the Japanese Empire, Figure 4.6.

(http://en.wikipedia.org/wiki/Japanese occupation of Korea)

Figure 4.6. Korea as Part of the Japanese Empire in 1939.

It was in this context of Korean society, in which the future leaders of Korea would grow up. In particular, we will review the origins of Park Chung-hee (President of the Second Republic of South Korea), Chung Ju-yung (founder of Hyundai), Lee Byung-chull (founder of Samsung), Park Tae-joon (builder of Posco), Syngman Rhee (President of the First Republic of Korea), and Kim Il-sung (Eternal President of North Korea).

Early Life – Park Chung-hee (President of the Second Republic of Korea)

Park Chung-hee was born in 1917 in the Japanese-occupied Korea. His father, Park Song-bin, had been of the Korean aristocracy (yangban) but was deprived of his birthright. He had participated in a ‘Tonghak’ rebellion in 1894 against the Korean monarchy and was sentenced to death (which was commuted). But Park's grandfather, fearing government retribution, disinherited this eldest son; and Park Song-bin had to move to land owned by his wife's uncle. He then existed in poverty, supported mostly by his wife's agricultural efforts. In the isolated countryside to gain an education, the young Park Chung-hee had to walk eight kilometers each day, to and from elementary school in the neighboring village of Kumi. The young Park persevered, showing the dedication and perseverance and ambition which were a mark of his character all his life.

Paek Nam-ui, Park's mother, endured the hardship of the poverty and supported her son. Lee Chong Sik Lee wrote: "Park's mother . . . had grown up in a relatively well-to-do family and experienced no hardships before her marriage. She was even able to read and write Korean script, a skill very few village women possessed. . . . she poured affection on her youngest son . . . she woke him early each morning with breakfast and a wash of warm water ready; and she stood at the edge of the pine forest to watch her son leaving for the two hour walk to school before sunrise and waited for his return after dark; she washed his muddy cotton-padded socks every night and laid them on the heated floor to dry for the next day; and she wrapped eggs in socks for him to carry to the Japanese-owned store by the school to exchange for pencils or notebooks." (Lee, 2012)

From such an early beginning, Park's ambition was to be a soldier. One of his biographers, Lee Chong-sik, who was fourteen years younger than Park, had shared the similar educational background as Park. Lee wrote: " . . . militarism was firmly established in the educational system in Korean when he (Park) attended . . . . Since Japanese was the only language permitted at school at the time, both of us (Park and biographer Lee) became fluent in it. Both of us read the war-oriented manga (comics), magazines and history books. We sang Japanese military songs and recited Imperial Rescript on Education. . . . the schools taught and trained us to be perfect subjects of Emperor Hirohito. Parents and family members did not dare to interfere in this process even if they disagreed." (Lee, 2012)

Park did well at the elementary school and was next admitted to the Taegu Normal School. This was the third school established in Korea by the colonial government to train teachers for Korea. His ranking in 1932 at admission was as 51st of the 100 students admitted, out of the 1070 students who applied. (Lee, 2012). But 1932 to 1937 was a time of depression throughout the world, and it was hard for his family to support his tuition. In 1936, the last year at the Normal school, Park's father arranged for him to marry a girl from a nearby village. Park was resistant because he had fallen in love with another girl, but Park acquiesced to his father's wishes. A daughter would result, but Park and his wife never found compatibility. Then upon graduation, Park was assigned to teach in Munkyong, a mining town near his hometown of Kumi, and he taught there from 1937-1940.

As we earlier noted, Japan had conquered Manchuria in 1931, re-naming it 'Manchukuo'. In 1937, Japan invaded China, swiftly conquering Nanking, Canton and Wohan. Japan needed a lot of soldiers. In 1938, Japan began inducting Koreans into the Japanese army. In 1939, the Kanto-gun Japanese Army in Manchukuo needed more officers for its expanding army and established the Manchukuo Military Academy. Park Chung-hee, who then was twenty-one years old, applied for admission. Park still wanted to be a soldier. Lee wrote: " Why did Park Chung-hee decide to cast off his teaching career and opt for the Manchukuo Military Academy? The short answer was because it was the only way for him to fulfill his lifelong desire to become Rikugun taisho, or an army general." (Lee, 2012)

Manchuria was also attractive: "Chung-hee had spent nine days in 1935 in Manchuria, as part of a school trip. This was the land of scarlet sunset and an endless horizon filled with sorghum, corn, and soybean fields. The cities, Dalian, Fengtien, Siniing, Antugn bustled with commerce and industry. . . It was clearly a land of opportunity, contrasting with the silent world of Munkyong. And this land beckoned the Koreans, because the Japanese militarists who conquered Manchuria wanted the Koreans to migrate there. . . (Historically) a part of Manchuria in fact belonged to the ancient kingdom of Koguryo (37 BC to 668 AD). .. . Manchukuo's motto (Japanese propaganda) was . . . 'Harmony of Five Races', the Japanese, Koreans, Chinese, Manchus, and Mongols" (Lee, 2012)

Graduating from the Manchukuo Military Academy, Park could immediately become an officer in the Manchukuo army or could enter the Japanese Military Academy, from which he could become an officer in the Japanese Imperial Army. He had difficulty being admitted, because of his age (then 23 but under an official age limit of 19); yet he was admitted to the Manchukuo Academy. After the first two years in the Academy, Park went next to Zama Japan to complete his training at the Japanese Military Academy.

Park Chung-hee in Japanese military attire. (Source: http://en.wikipedia.org/wiki/Park_Chung-hee).

The Academy was located only twenty five miles southwest of Tokyo and that region provided cultural pleasure to Park. Also for Park, the military subjects were intriguing. He studied the art of warfare, including the campaign failures of the Russian, French, and German generals. And the Japanese army's moral standards of loyalty, duty and frugality appealed to Park. Lee Chong Sik wrote: "How was loyalty and frugality related to the notion of overcoming death? Life is ephemeral; death comes to everyone sooner or later. Therefore, one could spend his short life searching for sensual pleasures or accumulating wealth, but these would not serve any purpose after death. The alternative would be to die for a cause . . . A person willing to sacrifice himself for a great cause would not absorb himself in material wealth or sensual pleasures because such acts would betray his insincerity." (Lee, 2012) This is how the biographer, Lee Chong Sik, explained Park Chung Hee's later dedication to the 'cause' of South Korean independence and economic growth and military might.

Given the brutal Japanese occupation of Korea for forty years and Park Chung Hee's ambition to become a loyal Japanese officer, it has puzzled many of Park's biographers how Park Chung-hee could have eventually turned out to be a really great South Korea leader – leading his country out of economic poverty, military weakness, political corruption – up toward becoming a global competitor. Lee wrote: "Park, who ruled Korea for eighteen years and supervised the transactions of billions of U.S. dollars, did not leave his heirs any inheritance worth mentioning. The efforts of muckraking press after his death led nowhere, and a consensus emerged that Park was a 'clean man'. This was in stark contrast to his five successors as Korean president, who were either imprisoned for corruption or saw their sons, brothers, or underlings sent to prison for the same reason." (Lee, 2012)

On 22 April 1944, Park Chung-hee graduated and was sent to Manchuria, as a second lieutenant in the Manchukuo Army. He was stationed in a remote part of the northwest, near Inner Mongolia and assigned to a small village Banbishan on the Xiao Luan River, accessible only by horse or donkey. There was little fighting in 1944, since by then the Japanese Army had defeated most of the guerilla forces in the region. He was isolated there for fourteen months. Finally on 17 September 1945, his unit learned of the surrender of Japan. Chong-sik Lee wrote: "The Japanese surrender on August 15 was like a bolt of lightning in a blue sky for the Eighth Regiment. . . the new situation changed everything for the Koreans. The Chinese officers and soldiers in the Manchukuo army now assumed the role of victors against the Japan's empire and disarmed the Japanese and Korean officers. . ." (Lee, 2012)

Along with two other Korean officers, Park followed the regiment southwest toward Beijing. In Beijing, thousands of Koreans waited to return to Korea. On 7 September 1945, the U.S. General MacArthur appointed Lieutenant General John R. Hodge to administer Korean affairs. The following May 1946, a ship took the Korean soldiers from Beijing to Seoul. Lee wrote: "His (Park's) wings clipped, a dispirited and exhausted Park Chung-hee returned to Korea in May 1946, only to face further hardship." (Lee, 2012)

We can see how the outlook and capabilities of a future leader of South Korea was influenced by his youth in Japanese-occupied Korea.

Early Life – Lee Byung-chull (Founder of Samsung)

Lee Byung-chull was born in 1910 to a wealthy, landowning yangban family, of the Gyeongju Lee clan. At the age of ten, Byung-chull was sent to a primary school in Seoul. We recall, in the 1930s, he studied in Waseda University in Japan (then a Japanese university which admitted Koreans); but he did not complete his degree. In 1936, his father died, and Lee Byung-chull used inherited money to start a rice mill in Daegu. It was not very profitable, and Lee moved to Taegu and began a trucking business and a real estate business, which he named 'Samsung', meaning 'three stars'. His business went bankrupt during a credit squeeze after Japan invaded Manchuria. Lee began his business again, building it. By 1945, Samsung Trading Company was one of the largest trading companies in Korea.

When the Korean War began, Lee escaped from Seoul to Pusan and continued to transport goods. In 1953 Lee built a sugar refining plant, called Cheil Sugar, First Sugar. In 1954, Lee established Cheil Wool Textile Company.

Early Life—Park Tae-joon (President of POSCO)

Park Tae-joon was born in 1927 in Yimrangli in Kyungsang Province. He moved to Japan when six years old, where his father was working. In 1945, he enrolled in Waseda University's College of Science and Engineering. But the war ended in 1946, and Park Tae-joon returned to Korea joined the military, graduating in 1948 from the Korea Military Academy. He studied under Park Chung-hee and served as an officer in the Korean War.

Early Life – Chung Ju-yung (Founder of Hyundai)

The founder of Hyundai, Chung Ju-yung was born in 1915 in a village near Tongchon (which is now in North Korea). His was a peasant family who worked hard to gain a living on marginal land. Peasant life was barely survivable, and Ju-Yung early decided to become other than a farmer, at first a teacher. But the family did not have the resources to fund his education beyond elementary level. His father needed all the childrens' labor to maintain the farm, but Ju-yung decided to escape from farming. At the age of sixteen, Ju-yung and a friend ran away at night, planning to walk to Chongjin, a city 250 miles away.

They made it to the town of Kowon, where they found work at a railroad construction site. The work was hard and long, for which they earned 45 jun per day. They found lodging in a dormitory, at the rate of 30 jun a day. But when it rained, there was no work, and costs were still 30 jun for the day. Also they had to replace clothes, frequently torn in the hard conditions. Chung Jun-yung quickly learned that a laborer's life in the city was not better than a farmer's life in the country. Then his father, after leaning where his son had fled, found him and asked him to return. Chung Jun-yung reluctantly returned.

The next spring in April 1933, Jun-jung again escaped from peasant life, this time to Seoul. But only ten days on the road, his father again found him; and he returned to help with the spring planting. After the harvest season, Jun-jung again escaped, taking 70 won, his father had received from selling a cow. Jun-yung took a train to Seoul and there enrolled in the Doksu Bookkeeping School to learn accounting. Two months later, his father once again found him, insisting he return to the farm. He argued that he needed four more months to complete his business course; but his father would not let him finish. Chung Jun-Yung worked hard for another year at the family farm, but the next spring an early frost destroyed the family crop.

Finally in 1938, Chung left again for Seoul. He worked first as a construction laborer, then as a handyman in a starch syrup factory. After a year, he gained a job as a delivery man for Bokheung Rice Store. He worked hard and helped to improve the operations and became the manager of the store. He had a modest salary, three meals a day and a half sack of rice per day. In January 1936, Chung returned to his village to marry a young girl, his father had chosen for him. The new couple returned to Seoul, and his wife helped the family income by making kimchi, dwenjang and gohujang to sell to customers, along with the rice. In 1937, the owner became ill and sold the business to Chung. Chung had his five brothers and one sister move to Seoul and help him with the family's 'new' business. But in 1939, the Japanese occupation outlawed Korean ownership of agricultural businesses, and Chung lost his shop, without compensation. He returned to his village and with some of his savings, bought additional land for his father.

A year later in 1940, he returned to Seoul, to start again. But now he was an experienced businessman, no longer a peasant. As a businessman, he knew to look for an available business opportunity and found one in a repair service business. Only Japanese were allowed to own automobiles, but Koreans could repair them. Chung bought a garage, which had employees who repaired cars. Chung would now learn the business of repair. Borrowing capital from a rice mill owner, with whom he had previously done business, he established his business as the Ado Service Garage. His business strategy was to repair cars more quickly than competitors. For four years, he operated a profitable business. But in 1943 once again, Japanese officials intervened, forcing him to merge his repair business with a Japanese owned local steel plant. Still he now had 50,000 won in savings, and he returned to his home village in Asan. After Japan lost the war, two years later in 1945, Chung could return to Seoul and to business.

Early Life – Rhee Syngman (President of the First Republic of Korea)

In 1875, Rhee Syngman (known in the West as Syngman Rhee) was born to a yangban family in Hwanghae Province in the Kingdom of Korea. His family moved to Seoul, and Syngman was educated in classic Chinese literature. But he applied for and failed civil service examinations. Next Rhee enrolled in a missionary school and learned English. In 1896, Rhee joined a political reform movement, Independence Club. They protested Japanese dominance, and Rhee was arrested and sentenced to life imprisonment. While in prison, he converted to Christianity and wrote a political manifesto, the Spirit of Independence.

When the Russo-Japanese War began in 1904, Korean government policy changed, and Rhee was released from prison. He was invited by U.S. government officials to a peace conference about the Russo-Japanese War; and he met with the U.S. Secretary of State John Hay and with the U.S. President Theodore Roosevelt. He pleaded, but without success, for U.S. intervention to keep Korea independent. In November 1905, Korea became a protectorate of Japan.

With assistance from missionaries, Rhee remained in the United States for an education. In 1907, he obtained a bachelor's degree from George Washington University and a master's degree and a PhD from Princeton University in 1910. His studies included politics, history, international relations, Christian theology and law.

In late 1910, Rhee returned to Korea as chief secretary to the Young Men's Christian Association. But in that year, Japan formally annexed Korea and began repressing the Korean Christian community. Rhee left for Hawaii and began a Christian school for Korean immigrants.

In 1919, Korean exiles met in Shanghai, China, to form a Provisional Government of the Republic of Korea; and Rhee was elected president of the government in exile. But in 1925, the organization removed him, after he was impeached by the Provisional Assembly for misuse of funds.

Rhee continued living in exile. He lived in New York, Washington, D.C., and Hawaii, where large Korean communities in exile were politically active. In 1933, Rhee met Franziska Donner, a native of Vienna, working in the League of Nations in Geneva. They married in New York in 1934, a second marriage for both.

After the defeat of Japan in 1945, the U.S. Army flew Rhee to Tokyo to meet General McArthur, then governing the conquered Japan. And in October 1945, McArthur flew Rhee to Seoul. Rhee was back in Korea.

Early Life – Kim Il-sung (Eternal President of North Korea)

Kim Il-sung was born in 1912 in a small mountain village near Pyongang and named Kim Song-ju. His father and mother were Presbyterian Christians and strongly nationalistic. In 1920, the Kim family fled to Manchuria, escaping Japanese persecution and famine. In 1926, Kim Il-sung attended Whasung Military Academy for a year and transferred to the Yuwen Middle School in Jilin, Manchuria. In 1931, he joined the Communist Party of China, as that year the Japanese Army invaded Manchuria. In 1935, Kim joined the Northeast Anti-Japanese United Army, which was a guerrilla group led by the Chinese communist party. Then Kim took the name of Kim Il-sung. In 1937 at the age of 24, he became commander of the 6th division. The Japanese pursued the guerrilla army. By 1940, Kim was a last Korean guerrilla leader still alive. Kim escaped Japanese pursuit by crossing the Amur River into Russian Soviet territory. He remained in Russia in a camp near Khabarovsk until 1945. The Soviets enlisted Kim as a major in the Soviet Red Army, and he remained in Russia until the Japanese surrender.

Korea Society and Korean Individuals

In the time-line of modern Korea, we have briefly traced the history from 1905 to1945, to the event of the end of World War II. And we have seen how this context influenced the early lives of the post-war leaders of Korea. The reason we have done this was to use the discipline of sociology (in addition to history) to understand the development of modern Korea. History is an empirical record of society, what happened and how it happened. But why did it happen? Why do societal histories occur? This is the explanatory domain of the social sciences. Thus we must expand our expository perspective from not only 'what' to 'why'? For the 'why' of historical explanation, we need to analyze history within a cross-disciplinary social science framework and the theory of societal dynamics provides this analytical framework.

Explanation of societal happenings is complex, usually requiring several explanations in a historical event. The concept of 'explanation' is basic to all science, physical and social sciences. However, there is a basic methodological difference in explanation between the physical and social sciences. The physical sciences explain changes (dynamics of events) in terms of cause-and-effect (causality); but the social science do not explain in cause-effect. To see this, one can see how the physical sciences view (observe) nature, in contrast to how the social sciences see, view, observe nature. Figure 4.7 pictures how a physical scientist observes nature in a physical space-time framework. The description of physical existence is that all physical objects exist in space and time, wherein anything is observed as occupying a position in the physical space (x,y,z) at a time t.

A three-dimensional geometric space X,Y,Z in which any physical object is described at a given time (t) at residing at a point described by the three coordinates (x,y,z). Objects can change positions in space (x,y,z) through motion; and so time (t) is used formally as a fourth dimension to enable the description of motion. A physical object can move from position to position over time. If a force impacts the physical object, its motion is can accelerated. This force is said to 'cause' the change (acceleration) in the motion of the object.

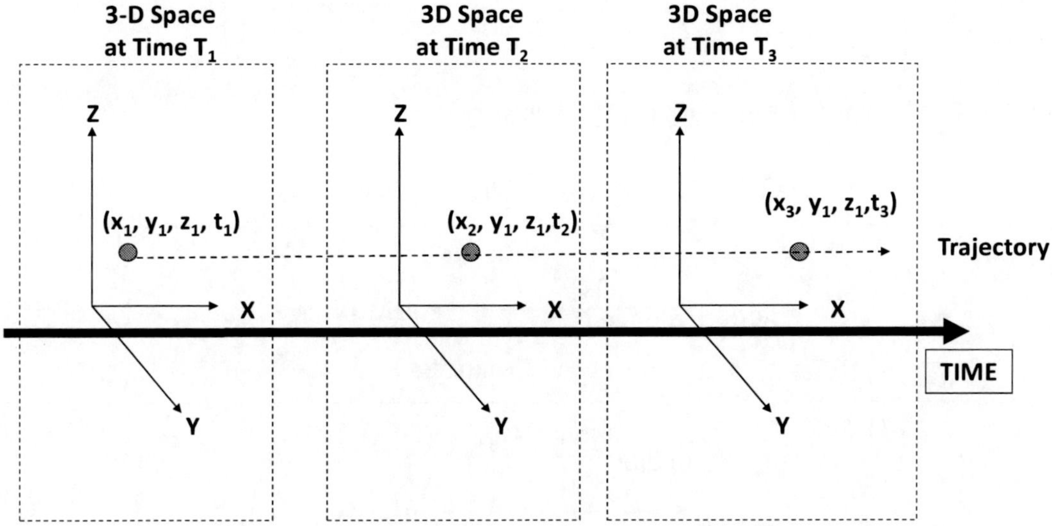

Figure 4.7. Classical Four-Dimensional Space-Time Description of Motion of Material Object.

In the physical-biological scientific disciplines, 'causality' is in the form of a force with alters the motion of a physical object.

But in historical studies of society (such as an economics history) one does not observe 'physical events' but instead 'societal events'. Thus physical concepts (i.e., 'physical objects', 'space and time', 'motion', forces', 'causality') are not useful (not methodologically appropriate) to describe societal history. Accordingly, there never is any 'causality' in historical explanation. But there are other kinds of explanations in history. What are they? To find them, we can use a methodological analogy to a 'physical observational space' – a 'societal observational space'. This has been constructed from the three basic dichotomies in the social sciences: *individual-society, groups-processes, and reason-action. (Betz, 2011)*

> The first basic idea in the sociological literature is how one can distinguish between individuals and the society in which the individuals live – the dichotomy of *individual & society*.
>
> The second basic idea in sociology distinguishes within a society how individuals associate into groups within a society and the processes a group inculcates in members – the dichotomy of *group & process*. A social process is a series of actions coordinated to produce an outcome planned by a group. In describing societies, sociological literatures use the basic descriptors of groups and organizations as societal things and of processes (social processes) performed by groups or organizations, such as political parties or corporations.
>
> The third basic idea found in the sociological literature (and in the management science literature) is about the thinking of individuals and their behavior. Individuals described as sentient (or cognitive) beings acting according to perceived reasons – the dichotomy of *action & reason*.

We can graphically show these three basic social science dichotomies upon a three-dimensional societal space, as shown in Figure 4.8.

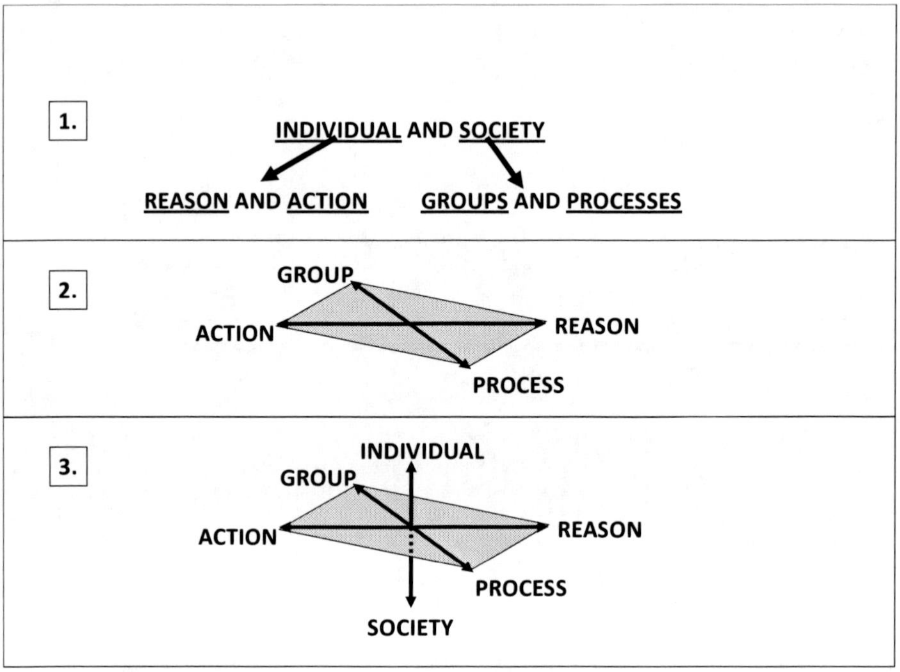

Figure 4.8. Basic Sociological Dichotomies.

In any historical event describing an epoch of a society, the event can be described as factors and interactions of three dichotomies (in the sociological perception of the event). These dichotomies are individuals-societies and action-rationality and groups-processes.

What we have been doing is describing the state of Korean society under the Japanese occupation and how some key individuals growing up in this society were influenced by societal conditions, and this is analyzed in the societal dynamics format in Figure 4.9.

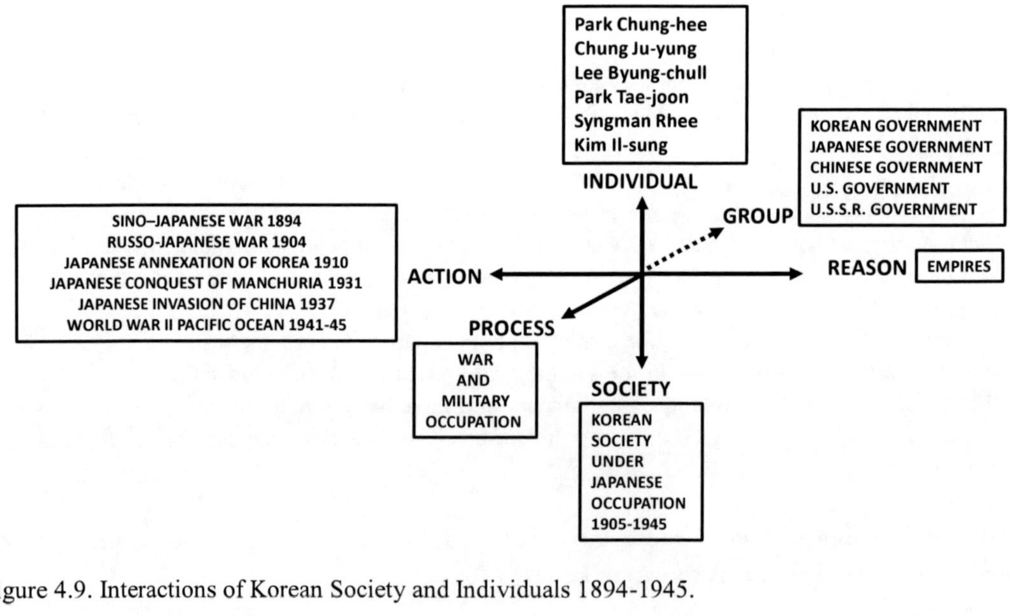

Figure 4.9. Interactions of Korean Society and Individuals 1894-1945.

Thus what we have reviewed is partial explanations of how future Korean leaders' characters were influenced by the social environment of occupation. Korean Society was shaped by actions (wars and occupation) in which different groups (governments) were involved in the processes of war and military occupations. And the reason underlying all these actions was the desire for 'empire' – Japanese empire to replace in Asia, the Western empires – British Empire, Russian Empire, French Empire, German Empire. We have done this to understand how leadership played an important role in the economic growth of South Korea. The conditions of pre-war Korean society, dominated by Japanese occupation, helped shape the characters of key individuals who, after the war, were to lead Korean society in independence and economic growth.

HISTORY: 1945 TO 1953 KOREA

Next we briefly review the history of the division of Korea and the Korean war. During World War II, the leaders of the Allies met in a series of conferences, Cairo in November 1943, Yalta in February 1945, and Potsdam in August 1945, Figure 4.10.

Generalissimo Chiang Kai-shek, Franklin D. Roosevelt, and Winsto Churchill at the Cairo Conference, 25 November 1943.
http://en.wikipedia.org/wiki/Cairo_Conference

Yalta Conference in February 1945 with (from left to right) Winston Churchill, Franklin D. Roosevelt and Joseph Stalin. Also present are USSR Foreign Minister Vyacheslav Molotov (far left); Field Marshal Alan Brooke, Admiral of the Fleet Sir Andrew Cunningham, RN, Marshal of the RAF Sir Charles Portal, RAF, (standing behind Churchill); George Marshall, Army Chief of Staff and Fleet Admiral William D. Leahy, USN, (standing behind Roosevelt).
http://en.wikipedia.org/wiki/Yalta_Conference

Clement Attlee, Harry Truman and Joseph Stalin at the Potsdam Conference, circa 28 July -- 1 August 1945
http://en.wikipedia.org/wiki/Potsdam_Conference

Figure 4.10. World War II Allied Conferences 1943-45.

Principle decisions included a demand for unconditional surrender of the Axis governments (Italian, German, and Japanese), Japan should lose all conquered colonies, and the Russians would declare war on Japan three months after the surrender of Germany.

Thus in 1945, after the Russian troops entered Manchuria and Korea, they stopped at the 38th parallel in Korea. This became the demarcation between U.S. and U.S.S.R occupation zones, Figure 4.11. The last Japanese Governor of Korea prepared to hand over the government of Korea in August to a group of Koreans. But on 7 September 1945, the U.S. General MacArthur appointed Lieutenant General John R. Hodge to administer Korea, and Hodge landed in Incheon with American troops. The Koreans calling themselves, the Provisional Government of the Republic of Korea, went to meet Hodge; but he refused a meeting. Instead Hodge, following U.S. orders, recognized Syngman Rhee as head of a provisional government. In February 1946, the U.S.S.R. occupied northern half of Korea and established a North Korean Provisional Government. Stalin appointed Kim Il-sung as head of North Korea.

Soviet troops in Korea, October 1945
http://en.wikipedia.org/wiki/Korean_War

U.S. troops in Korea, September 1945
http://en.wikipedia.org/wiki/Korean_War

Figure 4.11. Russian and American Troops Occupying Korea North and South of the 38th Parallel.

In November 14, 1947, the new United Nations passed a resolution that free elections should be held in Korea. But the Soviet Union did not consider the resolution binding. Elections were held in South Korea in April 1948, and Sygnman Rhee was elected president of the Republic of Korea. In the north, Stalin installed Kim Il-sung as prime minister of the Democratic People's Republic of Korea. Korea was henceforth divided at the 38th parallel.

In North Korea, Kim exterminated opposition and established his dictatorship of the communist government. Stalin provided military aid and funds to rapidly build a North Korean Army. In South Korea, some Koreans with communist leanings, refused to recognize Rhee's government and were brutally put down, with the uprisings lasting until the Korean War.

In 1950, Kim Il-sung travelled to Moscow, seeking support for an invasion of South Korea. Stalin approved of the plan but made it clear that Soviet troops would not engage. If troop assistance were needed, the Chinese would have to provide troops. Soviet generals were sent to advise Kim and plans for an invasion were prepared by May 1950. On 25 June 1950, the North Korean army crossed the 38th parallel. A tank brigade of 120 Soviet T34-85 tanks spearheaded the invasion and quickly overcame South Korean troops, unprepared for a tank invasion. On 27 June, Sigman Rhee's government fled Seoul. Figure 4.12.

The T-34 tank was standard armor used by the North Korean Army in 1950 and was also present at Osan. http://en.wikipedia.org /wiki/Battle_of_Osan

Hundreds of thousands of South Koreans fled south in mid-1950 after the North Korean army invaded. http://en.wikipedia.org/wiki/Korean_War

Figure 4.12. North Korean Army Overruns Seoul.

The North Korean Army had 40 soviet YAK fighters, and the South Korean army had no fighter aircraft. The North Korean Army had 150 Russian T34 tanks, the South Korean Army had none. The South Korean Army broke, and Seoul was taken within three days by the North Korean Army.

On the day of the invasion, the President of the United States Harry S. Truman was informed of the invasion. President Truman announced that the U.S. would counter the aggression and ordered General MacArthur in Japan to transfer material and reinforcements to Korea. President Truman sent an ambassador to the United Nations and obtained a mandate to assemble a United Nations Force to repel the Korean invasion. Earlier the Soviet Union delegation had walked out of the Security Council, thereby being unable to veto the resolution.

On 5 July, U.S. army forces in Korea engaged the North Korean Army at Osan, just south of Seoul. It was an infantry and artillery battery, but it also lacked tanks and effective anti-tank weaponry to stop the North Korean tanks. Driving their powerful Soviet T-34 tanks, a North Korean tank column quickly broke the U.S. army position. The remnants of the South Korean army and the U.S. army retreated south to Pusan. Neither the South Korean Army nor the resident forces of the U.S. Army had anticipated the invasion nor had prepared proper tank and air power to stop an invasion.

By the end of August, the South Korean and American troops, which had escaped capture collected behind the Naktong River, north of Pusan and established a defensive parameter. One 1 September 1950, The North Korean army launched an offensive against this 'Pusan Perimeter' of the South Korean and U.S. armies, see Figure 4.13.

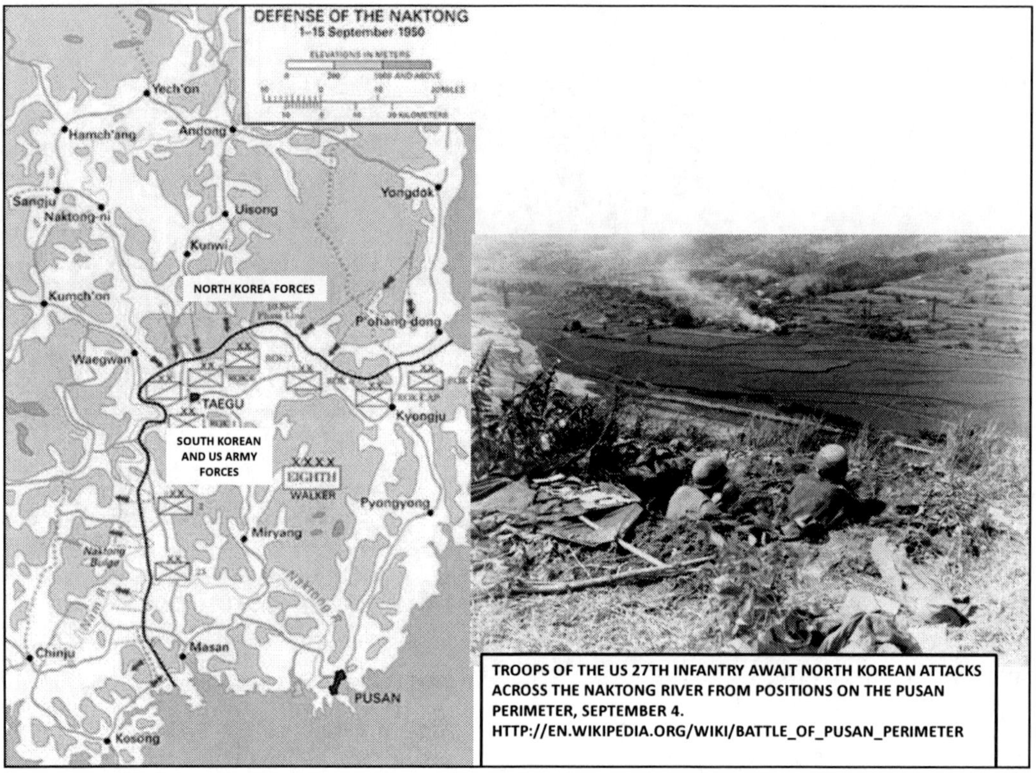

Figure 4.13. Korean War the Defense of Pusan at the Naktong River 1 September 1950.

The North Korean offensive tried for two weeks to break through the defensive line but failed. The Pusan perimeter held. Meanwhile, the U.S. General MacArthur planned a counterstrike. It was a classic military maneuver called a 'flanking attack' – hitting an army from the side. He planned to have American forces land at Incheon, on the coast near Seoul, and cut off the entire North Korean Army from its supply base in North China. The landing occurred on 15 September 1950, and the U.S. Army recaptured Seoul cutting off the North Korean Army, see Figure 4.14.

From the Pusan parameter, the southern U.S. Army unit attacked the North Korean army. The North Koreans began a battered retreat. About 40,000 North Korean soldiers were killed or captured, while another 30,000 soldiers managed to escape back to the north

On the 1st of October 1950, the Republic of Korea (ROK) and UN forces pursued the North Korean army north across the 38th parallel, up toward the Yalu River, the border between China and North Korea. On the 25th of October, 200,000 Chinese troops crossed the Yalu River began attacking UN forces. The UN and ROK forces had not anticipated the Chinese intervention and were thrown back, back down to the 38th parallel, see Figure 4.15.

FOUR TANK LANDING SHIPS UNLOAD MEN AND EQUIPMENT ON RED BEACH ONE DAY AFTER THE AMPHIBIOUS LANDINGS ON INCHON (HTTP://EN.WIKIPEDIA.ORG/WIKI/BATTLE_OF_INCHON)

GEN. DOUGLAS MACARTHUR (CENTER), COMMANDER IN CHIEF OF U.N. FORCES OBSERVES THE SHELLING OF LIGHTLY DEFENDED INCHEON FROM THE USS MOUNT MCKINLEY, SEPTEMBER 15, 1950.

Figure 4.14. U.S. Army Invasion of Inchon 15 September 1950.

Three commanders of Peoples Volunteer Army (PVA) during the Korean War. From left to right: Chen Geng (1952), Peng Dehuai (1950–1952) and Deng Hua (1952–1953)

Chinese forces cross the Yalu River.

http://en.wikipedia.org/wiki/Korean_War

Figure 4.15. China Intervenes in the Korean War.

The Korean War continued through the next year of 1951 and was largely a stalemate over territory. There were many fierce battles. The U.S. controlled the air and extensively bombed north Korean territory. But the bombing could not dislodge the half million Chinese soldiers holding the territory. Armistice negotiations finally resulted in peace, two years later in July 1953. The separation of Korea into North and South was permanent, for the next six decades.

LEADERS OF NORTH KOREA AND OF SOUTH KOREA

We noted that Kim Il-sung became the leader of the divided North Korea, because of his appointment to head the Korean Communist Party. After the Soviet army entered Pyongyang on 15 August 1945, Stalin needed someone to head the communist occupation of North Korea. Earlier, Stalin's head of the police, Lavrentiy Beria, had met Kim and recommended him to Stalin. The Russians sent Kim to Korea, who arrived on 22 August, as head of the Korean Communist Party. Kim established the Korean People's Army, and Soviet officers provided training and weaponry. In 1950, Kim requested permission from Stalin to invade South Korea, and Stalin agreed, except he would not send Soviet troops. After an initial successful invasion and then a subsequent rout, Mao sent in Chinese troops to save Kim's government.

Also we noted that the U.S. General MacArthur reinstalled Synman Rhee as President of the Republic of Korea, after recapturing Seoul in October of 1950. However in the Korean War, both northern and southern territories had been devastated. And reconstruction did not began until after the armistice in 1953.

In South Korea in 1952, Rhee had wished to be re-elected president. But the South Korean constitution at the time provided that the National Assembly elected the president; and the National Assembly was not pleased with Rhee. Rhee attempted to change the constitution for election by direct vote, and when the Assembly rejected this, Rhee had made mass arrests of his political opponents. Then Rhee had the amendment passed, and Rhee was elected president for a second term. In 1956, Rhee was elected for a third term, of which the constitution then allowed only three terms in office. In 1960, Rhee again had the legislature amend the constitution to a permit a president to be re-elected indefinitely, winning a fourth term, with 90% of the vote.

In 1960, the South Korean economy had not improved, and many viewed Rhee's government as corrupt. For example, Richard M. Steers wrote: "Shortly after the close of the Korean War, the Rhee government secured control of 166,000 different assets that had formerly belonged to the Japanese government or to Japanese residents of Korea prior to World War II. Initially held for safekeeping since 1945 by the U.S. occupation forces, these assets included 3,500 operating plants and companies, land, infrastructure, and inventories that accounted for close to 39% of the nation's wealth. Beginning in the early 1950s, officials of Rhee's government set about selling these assets, not to the highest bidders or the most qualified entrepreneurs and managers, but to those who paid the largest bribes to the government. . . At the same time, the Rhee government (through the Commercial Bank Act of 1954) began to privatize government-held banks, again selling not the the highest bidder but to friends and supporters." (Steers, 1999)

Also the U.S. had provided the main source of income to the South Korean government, and the U.S. support was declining. In 1960, Rhee was 80 years old and again desired to be re-elected and also to anoint his successor. He chose Gibung Lee to be elected as Vice President, and he received 8,225,000 votes against his opponent's tally of 1,850,000 votes. To many, this overwhelming ratio seemed fraudulent; and on 15 March, 1960, a protest occurred in Masan. Police shot at demonstrators. On April 11, the body of a student involved in the protests was found by a fisherman in the harbor. An autopsy disclosed that the student's skull have by split by a tear gas grenade. In Masan, mass protests against corruption broke out on April 19 and continued for three days, see Figure 4.16.

April Revolution
http://en.wikipedia.org/wiki/April_Revolution

Syngman Rhee in 1956
1st President of South Korea
http://en.wikipedia.org/wiki/Syngman_Rhee

Figure 4.16. April 1960 Protests.

On April 18 in Seoul, students from Korea University launched protests again police violence and demanded new elections. Protests continued through April 25. On April 26, Rhee resigned from power. The following day, his Vice-Presdient-elect committed suicide. On May 16, General Park Chung-hee exercised a political coup, and the Army took over the government.

Chung Ju-yung

In 1946, Chung Ju-yung returned to Seoul and restarted his automobile repair business; but he noted the construction companies earned more money. He started the Hyundai Civil Works Company and began doing construction projects for the Korean government. In 1950 when the North Korean army invaded, Chung fled to Pusan. His brother spoke English and acted as an interpreter for the U.S. army. Chung began contracting for construction projects of the U.S. army. When Seoul was recaptured in September 1950, Chung returned to the city and continued contracting for the army. He organized Hyundai Construction Company by merging Hyundai Motor Service and Hyundai Civil works Company. He built the new headquarters for the U.S. army in Seoul.

Chung saw that his U.S. army customers wanted projects completed on time and to be of excellent quality, caring less about the price. Richard Steers wrote: "Chung's strategy for success was simple. He concluded that, for the Americans, money was not the issue . . . . What the Americans wanted was guaranteed on-time performance in the middle of a chaotic, war-torn environment. . . . Looking back, it is clear that Chung and Hyundai received a great deal of assistance and support from their U.S. military contracts. . . They learned how to prepare project bids in the commonly accepted international format . . . Hyundai learned about Western specifications and was required to upgrade the quality of their construction work . . . They learned about construction management and quality-control techniques . . . they were able to acquire war surplus construction equipment that allowed them to mechanize their operations." (Steers, 1999)

Lee Byung-chul

We recall that in 1938, Lee Byung-chul had established a rice-trading business in Daegu, which continued until the end of World War II. In 1945, Lee expanded his business, by establishing the Samsung Trading Company for import-export of goods. His trucking business was transporting goods throughout Korea. When the North Korean Army invaded, He left Seoul to Busan. He continued his trucking business and also started a sugar refinery. In 1953, he opened an overseas trading office in Japan; and he also founded a food company, Cheil Jedang. Also in 1953, Lee founded Cheil Sugar (currently CJ Cheil Jedang). Using income from his successful businesses, he continued to found more businesses – in textiles (Cheil Wool Textile Co.), in cars, in insurance, in department stores (Shinsegae), and in consumer electronics.

Tony Michell wrote: "By the 1950s, Lee was already the richest man in Korea, and this following a decade of war in which Korea's rural landed wealth was destroyed. Lee was one of the new generation who learned how to profit from wartime and post-war situations and pursued his growing empire to acquire bank and insurance companies, as well as an increasing range of industrial entities. The majority of these were acquired rather than created. At the time Lee's philosophy for establishing a new business was to 'pounce on it when the market is at rock bottom'". (Michell, 2010).

Park Chung-hee

In May 1946, U.S. ships had moved Korean refugees from China to Pusan. Lee Chong Sik described that return: "While his LST was moored in the port of Pusan . . . Park Chung-hee would have noticed a new problem his country was facing. Pusan had been the busiest port in Korea, ferrying cargo between Japan and Korea. Ships carried away rice, fruits, farm animals, and other agricultural products Korea exported. Korea, in turn, imported manufactured products from Japan. . . . But in 1946, there was no trace of such traffic. All Park could see were the ships and LSTs loading the last of the Japanese returning to Japan and unloading Koreans returning from Japan and elsewhere. . . . The chaos in Pusan more or less epitomized the situation in postwar South Korea as a whole." (Lee, 2012)

Park found some of his old friends and learned that the U.S. Military Government was establishing a new training academy to train Korean police, Constabulary Academy. Park applied to the Academy and, after visiting his old home, he joined the Korean Constabulary.

Meanwhile, one of his brothers, Chung Sang-hee, was involved in a political movement, called People's Committees (which opposed the U.S. Military government). Earlier in February 1946, Stalin had set up the Provisional People's Committee for North Korea and placed Kim Il-sung in charge of North Korea. Kim Il-sung pushed for communists in South Korea to advocate integration of North and South Korea. The Korean Communist Party came to dominate South Korean political socialist movements and trade unions.

In August 25, 1946, the U.S. Military Government ordered the Korean Railroad company to reduce employment by 25% and turn employees into fewer full-time workers, instead of many part-time workers. The Korean Communist Party in the South encouraged the railroad labor union to strike on 23 September. Other unions then struck to form a national strike, such as printers union, postal and telephone workers. Rioting between workers and police broke out in Seoul and in Taegu.

The U.S. Military Government had continued the food rationing system, constructed by the Japanese, Local city officials and police forced farmers to give half their crops. This rationing system was behind the hatred of the People's Committees of the continuing U.S. Military Government. In Taegu in September 1946, people in a hamlet near the city of Taegu, entered the city and stormed the police station, killing police and burning records. Police from Seoul were sent in to put down the riot. In the battle, Park's brother, Chung Sang-hee, was killed by a police bullet.

Saddened by the death of his brother and angry at the U.S. Military Government for participating in putting down the People's Committee in Tague, Park joined an organization called 'Namnodang', Workers Party of South Korea. It was opposed to the formation a a separate South Korean State, and in February 1948, it instigated general strikes and a popular uprising on Jeju Island. In May 1948, elections were held to establish the Republic of Korea, and Syngman Rhee won a seat in the Assembly and then was elected President of South Korea on May 31. President Rhee outlawed the Workers Party of South Korea. He formed the ROK army and ordered all officers of the former South Korean Constabulary to be purged of Namnodang members.

Ten percent of the former Constabulary were arrested in September 1948. Park Chung-hee was also arrested and tortured and convicted of treason. On February 8, 1949, Park was sentenced to death. The new commandant of The Air Force Military Academy was a colleague of Park and appealed for his release. He knew Park was not a communist. Park's sentence was commuted to life imprisonment, then reduced to fifteen years and then suspended. Park returned to a position he had held as head of Operational Intelligence. But in April 1950, he was informed that he was dishonorably discharged from the Army. In June 1950, the North Korean army invaded, and Park was called back into the South Korean Army.

Park Chung-hee conducted an orderly retreat of his group from Seoul To Taejon, taking with them key maps and charts. The ROK army needed experienced officers. Park was assigned as an artillery corps commander and sent to the U.S. for artillery training. In November of 1953, Park was made brigadier general. In 1951, Park met and married Yuk Yong-su, and they had a daughter, Guen-hye.

On 16 April 1960, Syngman Rhee announced his resignation from a fourth term as President due to student demonstrations. On 16 May 1961, General Park took over the South Korean Government.

Park Tae-joon

Park Tae-joon (who as we earlier described, had founded Posco) had known Park Chung-hee as a student, when Park Chung Hee was an instructor in the Korean Military Academy. After the Korea War began in 1950, Park Tae-joon was promoted to the rank of Major, and in 1953 he was promoted to Chief of Staff of the Fifth Division of the Korean Army.

Kim Il-song

After the 1953 armistice, Kim Il-song was restored to leadership in North Korea. He began a national plan to rebuild North Korea, focusing upon heavy industry. The Soviet Union built factories and supplied materials and financial support. Kim Il-song collectivized agriculture, like the soviets; and everybody worked for state factories or farms. Chinese troops withdrew from Korean in 1957.

To solidify his power, Kim purged communist officials, sending victims and their families into prison camps. Kim's hold on power was absolute and brutal, a communist dictatorship, similar to the communist regime of the Soviet Union and China, his allies and supporters.

Explanation in a Societal Perceptual Space

We have analyzed history within the framework of a cross-disciplinary observational space of society, a societal perceptual space. This analytical perspective is methodologically useful in identifying the types of explanatory relationships, which can examine and understand how and why things happen in a society. The theory is called 'societal dynamics' and lists fifteen kinds of explanations which occur in historical events, as shown in Figure 4.17.

We will use these explanations to describe the impact of environment upon character of the future Korean leaders. This analysis of societal events in the three dimensional space of the societal perceptual (observational) space can be connected by fifteen lines in a topological graph. (Betz, 1012)

Each graph relationship provides a kind of explanation connecting two factors. (Topology is an area of mathematics concerned the connectedness of geometrical forms, and some topological graphs display this connectivity as lines connecting points in the geometrical form.) The societal dynamics event graph displays the explanations connecting principle factors in a historical event.

TOPOLOGICAL GRAPH OF EXPLANATORY RELATIONSHIPS

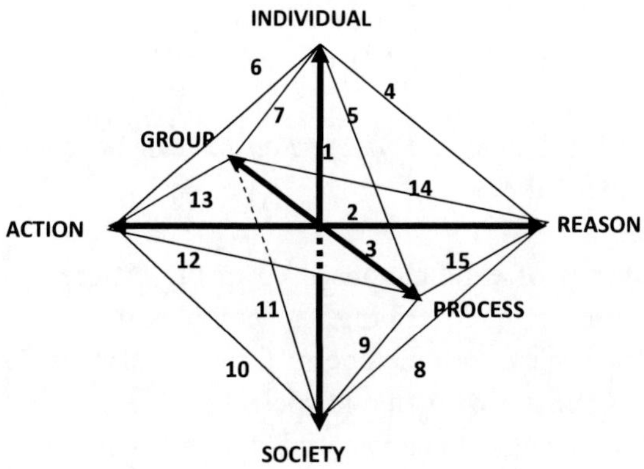

LIST OF EXPLANATORY RELATIONSHIPS

1. ETHICS
2. PRINCIPLES
3. INSTITUTIONALIZATION
4. IDEAS
5. POLICY
6. STRATEGY
7. GOVERNANCE
8. KNOWLEDGE
9. REGULATING
10. PERFORMANCE
11. INFRASTRUCTURE
12. TECHNOLOGY
13. OPERATIONS
14. IDEOLOGY
15. SYSTEM

Figure 4.17. Explanations in a Historical Societal Event as Relationships between the Principle Factors.

The fifteen kinds of explanations which can occur as relationships between the principle factors are:

1. ***Ethics*** – The explanatory relationship between *Individual-Society* can be called the *ethical context* in the explanation of an historical event.
2. ***Principles*** – The explanatory relationship between *Reason-Action* can be called the *principles-of-order* in the explanation of an historical event.
3. ***Institutionalization*** – The explanatory relationship between *Group-Process* can be called the *institutionalization* in the explanation of an historical event.
4. ***Ideas*** – The explanatory relationship between *Individual-Reason* can be called the *ideas* which an individual uses in reasoning.
5. ***Policies*** – The explanatory relationship between *Individual-Process* can be called the *policies* an individual in power formulates to control social processes.
6. ***Strategy*** – The explanatory relationship between *Individual-Action* can be called the *strategy* in which a leader formulates the direction for action.
7. ***Leadership*** – The explanatory relationship between *Individual-Group* can be called the *leadership* of an individual in guiding the efforts of a group.
8. ***Knowledge*** – The explanatory relationship between *Society-Reason* can be called the *knowledge* which a society has to use.
9. ***Regulating*** – The explanatory relationship between *Society-Process* can be called the *regulation* of activities within the infrastructure of the society.
10. ***Performance*** – The explanatory relationship between *Action-Society* can be called the *performance* attained by processes in societal sectors.
11. ***Infrastructure*** – The explanatory relationship between *Group-Society* can be called the social *infrastructure* which groups provide in building and operating sectors of a society.

12. ***Technology*** – The explanatory relationship between *Action-Process* can be called the *technology* used in a a process in producing an action.
13. ***Operations*** – The explanatory relationship between *Action-Group* can be called the *operations* of a the group which produce a group action.
14. ***Ideology*** – The explanatory relationship between *Group-Reason* can be called the **ideology** as the concepts groups use to associate and justify association, the ideology of a group.
15. ***System*** – The explanatory relationship between *Process-Reason* can be called the *system* of controlled process in the societal event.

This list is a cross-disciplinary in the kinds of explanations used in the different social science disciplines; and it is particularly useful in describing the complexity of societal events. It provides both a generality and a logical completeness, for comparing types of explanations across different historical events and across different societies.

Next we can analyze the history of events we have been examining about post-war Korea and the characters of Korean leaders in this analytical framework, Figure 4.18. The biographies of Korean leaders we have been looking at were Park Chung-hee (President of the Second Republic of South Korea), Chung Ju-yung (founder of Hyundai), Lee Byung-chull (founder of Samsung), Park Tae-joon, Rhee Syngman (President of the First Republic of Korea), and Kim Il-sung (Eternal President of North Korea). These are the 'INDIVIDUALS', as noted on the individual-society dimension of the societal perceptual space.

The relationship of such individuals directly to society is an 'Ethical' relationship. Society provides the ethical context in which individual characters are explained, and individual actions in society express real ethical behavior. And in all the biographies, we saw a common ethical commitment of 'patriotism'. Each individual was committed to love of their country.

But patriotism differed between the South Koreans and the North Korean Kim Il-sung. The reason was because of differing ideologies. Kim held the ideology of communism, versus the capitalistic ideology of the South Koreans. As a result of this difference in ideology, there was a difference about the form of 'governance'. Kim believed in 'dictatorship' – in the Lenin/Stalin/Mao dictatorship of the proletariat, meaning dictatorship under a communist party.

The ideology of capitalism, accompanied by a belief in democracy, was characteristic of the South Korean leaders. Still democracy was not a simple concept for the South Koreans. The democratic issue was about the election of the president – by whom and for how long and in what kind of elections. Democracy for Rhee Syngman meant his eligibility to be reelected indefinitely in rigged elections. Democracy for Park Chung-hee also meant continual re-election. Both believed they were essential leaders for South Korea at that time of contention with North Korea. Democracy with fair elections and term limits was an institution which South Korea would take some time to work out. But this evolution of economic and political ideas would never occur in North Korea.

Also in ethics, we saw a common desire for 'justice', since Koreans had experienced injustice under the Japanese occupation. For example, in Chung Ju-yung's experience, every time he started a successful company (rice business, auto repair business), Japanese officials would take it away without compensation. Byung-chull likely felt injustice during the Japanese occupation times whenever he had to pay bribes to remain in business. But the idea of 'justice' differed among the Korean leaders. For the North Korean leader, Kim Il-sung, justice was absolute loyalty to his person (personality cult of a communist dictator). For Rhee Syngman, justice was the give-away of state property to individuals supporting his campaigns; and injustice was any protest against his policies, which should be put down by police action.

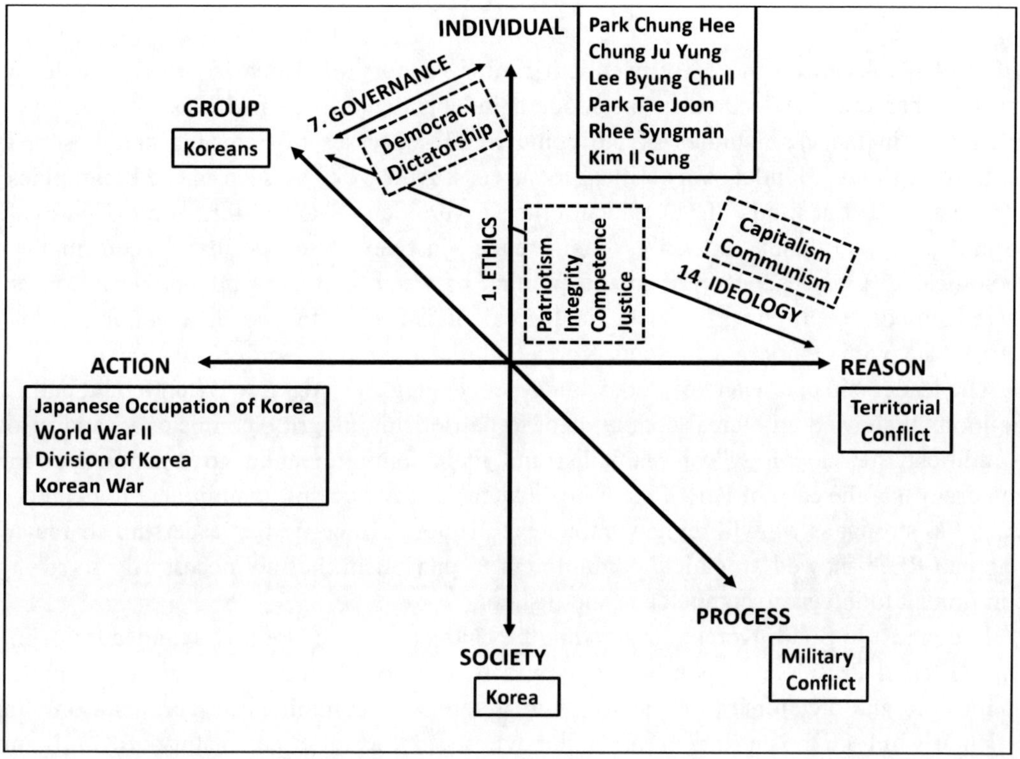

Figure 4.18. Perceptual Space Analysis.

'Integrity' was another ethical value experienced by the characters of the future Korean leaders. In Chung Ju-yung's experience, we saw how 'integrity' became the basis of his business competitiveness – his word was his bond. Also in Park Chung-hee's experience, he valued the integrity and honor of the soldier, a willingness to die for a cause. But integrity was not highly valued in Rhee Syngman's character, as there had been little integrity in his administration but instead much corruption. For the dictator Kim Il-sung, his idea of integrity under communism was loyalty to his person and absolute rule.

We saw the ethical value of 'competence' in Chung Ju-yung's capability to learn new businesses and the management of new businesses, e.g. auto repair and construction. Park Chung-hee highly valued the military competence of the Japanese army and distained the incompetence of the Korean emperors to defend their country from occupation. In an earlier chapter about the case of Samsung, we saw how Lee Byung-chull valued competence so highly that he funded concurrent R&D projects over a decade in order to catch up and surpass Japanese and Taiwanese competence in the production of DRAM chips. Also in an earlier chapter about Posco, we saw how Park Tae-joon was totally committed to competence in constructing the steel production facilities on time and under budget for profitable operations.

Neither Rhee Syngman nor Kim Il-sung much valued 'competence', as neither of their armies was successful in preventing nor succeeding in invasion. They both primarily valued 'loyalty' in their officers, over competence.

CONCLUSION

We have reviewed the early twentieth century history of Korea in order to understand how the characters of leaders in the modern independent Korea had been formed. As we earlier saw in the case studies of the commercial successes of Samsung and Posco, that leadership of their founders was critical to success. And we have seen in the biographies of early lives that four kinds of (1) Ethical criteria were central to the formation of individual characters – patriotism, integrity, competence, justice. Also we have seen that (14) Ideological differences occurred in the politics between southern and northern Koreas as 'capitalism' or 'communism'. Such ideological division led to two different forms of (7) Governance, with democracy in South Korea and dictatorship in North Korea.

The leadership characters of Park Chung-hee, Chung Ju-yung, Lee Byung-chull, and Park Tae-joon all showed an ethical commitment to patriotism, integrity, competence, and justice. In addition, their ideology was capitalist and their commitment to governance was for a democracy. (In the case of Park Chung-hee's later governance, his commitment to democracy was not as strong, as we will review in the next chapter). However, the leadership character of Syngman Rhee showed an ethical commitment to patriotism, but not necessarily so strong a commitment to integrity, competence and justice.

In contrast to South Korean leadership, the character of the North Korean leader, Kim Il-song ('Eternal President of North Korea) was focused only on ideology and governance – communism and dictatorship. Ethically, Kim Il-song was committed to patriotism only, and particularly only to the patriotism of a nation wholly to the cult of personality of Kim Il-sung. Ethically, Kim was incompetent in launching the invasion of South Korea, with U.S. troops present on the territory; and Kim had to be saved by the Chinese. Ethically, integrity and justice had no meaning within the ideology of a communist dictatorship.

Early life and one's social conscience can sometimes be tightly related. For example in the character of Park Chung-hee, at least one biographer, Lee Chong-sik, thought this so of Park: "The germ of Park's anger, or his sociopolitical consciousness, was planted very early in his life. The young child could not help but notice the wide discrepancy in wealth and lifestyle between his immediate family and that of his third uncle, his father's younger brother. His uncles' family could enjoy rice, the most valued and preferred staple food of the Koreans, while is family had to be content with barley or millet, the coarse and cheapest grain." (Lee, 2012) And the source of the poverty of this father, in contrast to the relative affluence, of his uncle, the father's younger brother, traced back to a societal event, the Tonghak movement – a connection between the *Individual,* Park Chung-hee, and *Society*, Korean history after 1894.

We recall that the emergence of modern Korea can be traced from that time in 1894 of the first Sino-Japanese War. Then Japan began exercising military influence over Korea; and this was a result of a Tonghak rebellion. In this rebellion, the Korean monarchy asked

Chinese help to suppress it; and after Chinese troops entered Seoul, the Japanese declared war on China.

Historical events – individuals – society. We recall that Park Chung-hee's father, Park Song-bin, was arrested for participating in the Tonghak movement. And to ensure the Park family safety, Park Song-bin had been disinherited by his father (Park Chung-hee's grandfather). Lee Chong-sik wrote: "The immediate cause of Park's poverty was his father's involvement in the Tonghak movement, which swept the country in the late nineteenth century. His grandfather had inherited enough land to feed the family. Having the inheritance of 'yangban', he naturally wanted his first son, Song-bin, to climb the ladder to success, which meant preparing for and taking civil service examinations. . . . Why did Park Song-bin join the Tonghaks? Why did the government condemn the rebels . . meting out a death sentence on Park Song-bin? . . . . It was sheer luck, as his wife told her son (Chung-hee) later that he had escaped the hands of the executioners. . . . (But) Park Song-bin had risked too much when he joined the Tonghaks. . . . Song-bin's father ousted him from the clan." (Lee, 2012)

We also recall that it was for this reason, Park Chung-hee had grown up in poverty on a small plot of mountainous land, which his mother had been allowed by her brother. The young Park Chung Hee had resented the treatment his family received as 'disinherited yangban. Partly due to this childhood experience, Chung-hee, all his life, felt deeply about justice for all the Korean people. Up until 1887, half of the peasants in Korea had been free but the other half were slaves of yangban. Foreign merchants coming into Korea increased the price of foodstuffs; but farmers did not benefit since yangban landowners collected half the crop as taxes. A severe famine occurred in Korea in 1892. Tso Sun-un organized an underground movement among common people against the Korean monarchy. In March 1894 in Ko-pu county, peasants began fighting government police, with the motto: "Save the people from the corrupted government." (Sunoo, 1994). This was the beginning of the Tonghak rebellion (which led to the first Sino-Japanese war in that year of 1894).

Sunoo Harold Hakwon wrote: "The Tonghak Party proclaimed: 'The highest thing in society is the ethics. . . the government officials do not think of the people's interest; on the contrary they bribe the ones in high official positions, waste their wisdom, condemn the honest scholars . . . there is not outstanding talent among the officials, foreigners crowd into the government circle. The resentment of the people against these things increases everyday . . . we cannot be patient any longer. We will fight to save the people, and will fight to the death." (Sunoo, 1994) This paragraph really gets to the heart of Park Chung-hee, to his character, as he emerged a soldier. Let us state again this Tonghak statement from 1894, from the time of his father:

"The highest thing in society is ethics *(Ethics)*

- government officials do not think of the people's interest; on the contrary they bribe the ones in high official positions *(Lack of Integrity, Corruption)*
- there is not outstanding talent among the officials (*Competence*)
- foreigners crowd into the government circle *(Patriotism)*
- we cannot be patient any longer *(Justice)*
- we will fight to save the people and will fight to the death *(Patriotism).*"

In 1961, General Park Chung-hee seized the South Korean government in a military coup. His motivation was for integrity, competence, patriotism, justice. This is history to which we next turn.

LESSONS LEARNED

Good leadership in a nation of its government and of its industries is essential to national economic growth, justice, and freedom in the nation.

Ethics is basic to any society; and leaders must have integrity and avoid corruption.

Leaders must also be competent is formulating strategy and in governing.

Leaders must also be patriotic for the good of the whole county.

Leaders should ensure that justice prevails in governance.

Leaders will defend a nation against its enemies.

Chapter 5

Industry and Government Partnership – Korea

Introduction

We have seen in the cases of Samsung and POSCO that leadership was an important factor in their commercial successes. The literature on economic development has not always emphasized the importance of the roles which particular leaders play in economic development. Instead a common assumption of economic theory has been that if some entrepreneurial individual had not appeared in economic history, another would have taken his/her place. This is a societal view of history. The social system provides opportunities for individuals to behave entrepreneurially in an economy. If the proper societal system (i.e. capitalism) does not exist, then capitalists are not possible. But always this is only partly true, because not all entrepreneurs are alike. Particular individuals do make a difference in history.

This was true of North and South Korea. In North Korea, capitalists were not allowed to exist (killed or imprisoned) under the communist regime of Kim Il-sung. But Kim Il-sung as the particular leader altered the direction of North Korean society. Yet a purely societal-theory view of history would ignore the historical difference a particular individual makes. What if Kim Il-sung had not been made leader of the Korean Communist Party, and some other communist had been selected by Stalin? Would an 'alternative-reality' communist leader of North Korea necessarily have invaded South Korea? Also could such an alternative-leader have managed to keep the dictatorial regime in place for so many years, with a son Kim Jon-il and a grandson Kim Jon-un?

Particular individuals do make historical differences through their particular personalities and decisions. Thus in analyzing societal histories, one needs the dichotomy of both *Individual and Society* to explain societal events. As particular individuals act in history, their characters matter in the determination of historical direction, yet there are general social forms in which they act. Societal contexts influence the development of individuals' characters; and particular individuals can alter the societal context.

In the case of Korean economic development, these societal forms were (1) chaebols and (2) government policy. Here we will examine how social forms and individuals' characters interacted in modern Korea – Society and Individual. For this reason, we previously reviewed the characters of Korea's post-war leaders, evolving when young in the context of an occupied Korea. After 1961, the character of Park Chung Hee played a definitive role in altering the history of South Korea.

Korean Chaebols

Even in a global economy, the forms of governance of a global company can differ. In Korea, a global firm is called a 'chaebol', from the term 'chare' (or 'wealth) and the term "pol' (clan). A US-based global company is governed as a 'holding company' (a company which holds or owns other companies). A Japanese-based global firm is organized as a 'keiretsu', which is a group of cooperating firms whose equity is partly owned by a bank (around which the keiretsu is organized). A Korean chaebol is a particular form of a group of companies controlled by a family. Chaebols have not been holding companies (one company owning other companies in a group); nor have they been keiretsu (or the earlier zabatzu), as a group of companies controlled though equities owned by a bank.

A Korean chaebol is a set of enterprises controlled by a single Korean family. How they came about was this way. Feudal societies are family and clan structured, and so businesses therein are family structured. Under the Japanese occupation, Korean owned businesses were structured around a family. For example, businesses in both Samsung and Hyundai had been started by Lee Byung Chull and Chung Ju Yung while under Japanese occupation. Of the fourteen largest chaebols in 1960, six had begun under the occupation and eight under Rhee's government. (Kang, 1996)

How did Korean enterprises survive under Japanese occupation? This was not easy. We saw in Chung's early biography, that each time, he started a successful business (first as a rice trader and then in automobile repair), Japanese officials confiscated his businesses. Some form of cooperation with Japanese officials and enterprises was necessary for Korean businesses to survive during the occupation.

About Korean businesses in the time of Japanese occupation, Kang Myung Hun wrote: "A distinctive pattern of investment portfolios and election to corporation boards was apparent by the last decade of the colonial period. The pattern included four concentric levels of business activity. First an inner core of family-owned companies, usually in agriculture and real estate. . . . The second level of investments included major investments with administrative responsibility in complex joint-stock enterprises. . . A third level included smaller investments in medium-sized ventures, without major administrative responsibilities . . . a team of major Korean entrepreneurs and a few Japanese government officials. A fourth level . . . was minor participation of Korean capital in . . . large Japanese investments in the peninsula." (Kang, 1996)

The protection of Korean business from confiscation by Japanese officials occurred when Korean entrepreneurs had investments in the third and forth level, providing capital to Japanese-led businesses. For this Japanese officials might protect a Korean partner. The first level originated from agricultural and real estate holdings, providing the capital base upon which a chaebol began.

In the case of Chung Ju Yung, we saw his family was peasant, without a yangban large agricultural holding. So while Chung participated in the first level of investment (rice business and auto repair business), he had not yet progressed to capital investments in

Japanese controlled enterprises (third and fourth level). Hence his businesses were not protected by Japanese officials and were confiscated.

The chaebols were family-controlled groups of businesses, both those begun under Japanese occupation and later under Rhee's government. These were the capital entrepreneurs which Park would have as partners to build the modern South Korea.

History: Park's Military Coup of 1961

We recall that on the 19th of April 1960, student protests in Seoul forced Syngman Rhee's resignation from his fourth term as president of the Republic of Korea. Students from Korea University marched to the presidential palace in the centre of Seoul. Soldiers fired on the students, killing 140 students and wounding many others. Protests continued on April 25th, with professors joining students and citizens. Rhee resigned. Prime Minister Yun Bo-seon formed a new government on August 14, 1960. But protests continued, over a poor economy and governmental corruption. The following spring on 16 May, 1961, Major-General Park formed a Military Revolutionary Committee and took control of the government.

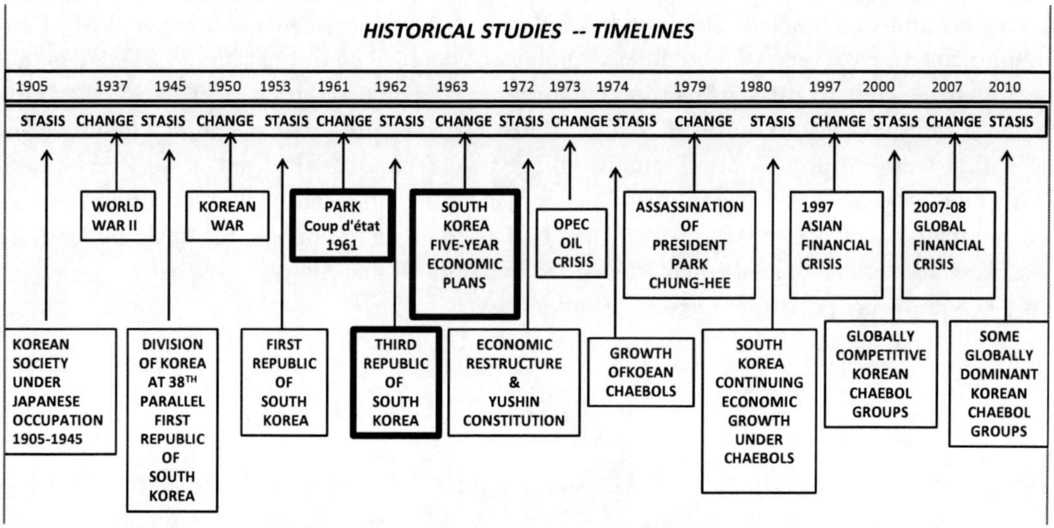

Figure 5.1. Change Events in Republic of South Korea.

Park Kisung described the coup: "Park's peers in the military supported him. The leading powers in the Park regime were military leaders. Many of them had crossed the border into Manchuria in the colonial eara and entered the Manchurian Military Academy. . . Park appointed his subordinates to important positions in the government. . . . (But) unlike Rhee's government, Park's government concentrated its power on the administration to prepare (for) . . . economic growth. . . . Park's government formed a powerful bureaucracy . . . appointed various specialists to government departments. . . The economic policies . . . were mainly swayed by the Economic Planning Board, (having) powerful authority over the allocation of resources . . Their influence on the overall economy of South Korea was tremendous." (Park, 2008)

In 1962, Park Chung-hee published his version of the coup, which he called a 'revolution': "As all my beloved people know, our Fatherland is now confronted with a decisive moment in its history; . . . We are standing at that important crossroad . . We are fighting an ordeal that no nation has ever overcome, unless it showed remarkable qualities of courage, wisdom and perseverance." (Park, 1962) Park's view of his military coup was that the nation of South Korea was in crisis and desperate political action was necessary.

In his book, to justify his revolution, he noted several revolutions of the past: "1. Modernization of China and Sun Yat-sen's Revolution; 2. Meiji Reform and Modernization of Japan; 3. Kemal Pasha (Ataturk) and Turkish Revolution; Nasser and Egyptian Revolution." (Park 1962) Thus Park saw his own action not merely as a political 'coup' but as a 'revolution' to save the county.

The Korean War had left Park's country in a perilous position. Although accused of having been a communist and sentenced to death under Rhee's regime, Park was never a communist. Park admired the spirit and competence of the Japanese army but not the communists. Park saw the country still under threat, after Rhee's incompetence and corruption, and still at war with communist North Korea.

Rhee's government had depended upon U.S. aid, and as the aid declined, the South Korean economy contracted. Thinking himself as a revolutionary (but not a communist), Park Chung-hee wrote: "Every year countless numbers of people had to fight hunger. The idea that a man has the right to eat and live is no luxury, no excessive ambition. . . The time has now come, however, when we must put a stop to this history of tragedy. . . The key industries of the nation were in pitiful state. Farming villages were impoverished and miserable beyond description. The crowd of unemployed intellectuals in the cities was a heart-breaking sight. . . We say the country was poor. . . more than half of the national budge, the basic budget, the housekeeping national management, depended upon the United States. . . . What would we do if the U.S. aid were cut off tomorrow?" (Park, 1961)

President of South Korea
In office
December 17, 1963 – October 26, 1979

President of South Korea (http://en.wikipedia.org/wiki/Park_Chung-hee).

This was how Park saw the urgency of the time and the failure of Rhee's government – an impoverished country, still at war and dependent upon a foreign country for its survival. In 1962, the U.S. pressured Park' military junta to restore democracy. Park ran for president, wining with a 46.6% vote over his opponent, Yun Po Son at 45.1%.

President Park's priority focused upon growing the Korean economy. In 1964, he established an Economic Planning Board (EPB) to control both the budget and policies of the government to stimulate economic growth. The EPB was established with technocrat professionals, and not political cronies. It stimulated economic growth through private capitalists and did not build state industries. Park's ideology was capitalism and not communism (the Japanese government after the Meiji restoration had been capitalist). Parks' South Korean government intended to use capitalism to build a strong military (as earlier had the Japanese). Park's ambition was to be a general in a strong, independent Korea. President's Park's economic model was probably influenced by his early experiences in Manchuria, wherein he had witnessed the economic growth in Japan's 'Manchuko', in a partnership between Japanese capitalists and the Japanese army.

Park's EPB was to work with the South Korean capitalists, the chaebols, to build the economy. From 1962 to 1967, Park's EPB launched its first five-year economic growth plan. Kang Myung Hun wrote: "Korea's sustained rapid economic growth began shortly after the establishment of a government headed by Park Chung-hee . . . The role of the government during the rapid growth of the chaebol was significant. . . . Korea had had one of the lowest income levels in the world, poor social and economic infrastructure, a small number of proven entrepreneurs and managers, a great information asymmetry (i.e., low level of technology), and a low level of domestic savings and technology. . . . (Under Park's government in) the 1960s, the Korean economy achieved an unprecedented rate of quantitative growth. In this process, many chaebol benefited from the government's allocation . . . to the private sector. . . . The difference (was) . . . during the Rhee period, there was dominantly the transfer of resources (e.g. properties and materials) . . . (whereas) during the Park period, there was a transfer of opportunities (e.g., license to invest)." (Kang, 1996)

After assuming government control, Park was determined to wipe out the previous corruption of the Rhee period – by punishing people with criminal sanctions. Lee Byung-chull (head of the Samsung chaebol) met with Park and offered a compromise. Bang Jung Hyun wrote: "President Park Chung-hee staged an anticorruption campaign in 1961. At the time, Lee was residing in Japan, refusing to go back to Korea and confront the incumbent president, because he did not want to walk into the mouse trap where he will be the main target as the richest man in Korea. Nonetheless, he and President Park made a deal that ultimately became the pillar for Korea's conglomerates. The deal was as follows: Samsung is allowed to keep the business running, but it has to be the carrier that implements governmental projects that are in concurrence with President Park's economic plan. . . . (This) established the nature of Korean economic system, in which the government and the business circle formed a common ground of interest that would benefit the two." (Bang, 2010)

President Park then implemented this compromise as an "Investment Law", in which criminal sanctions were removed from the chaebols and they were allowed to participate in major projects (specified by Park's EPB five year economic plans). Kang wrote: "In August 1961, these entrepreneurs formed the Korea Businessmen's Association, which became an influential interest group. The members also traveled abroad to solicit the foreign capital necessary to finance the projects. The association evolved into the Federation of Korean

Industries (FKI) in 1968 and gained influence ever since in the Korean economic scene." (Kang, 1996)

President Park's economic priorities were implemented through five-year economic plans. But what was an 'economic plan' for a capitalist economy? In capitalism, government does not own or run businesses. Private businesses plan their own enterprises. So what was Park's intention of a national economic plan? It was a plan for policies to assist private capital businesses to develop and contribute to national economic growth. This was the government-industry partnership to which General Park Chung-hee and Samsung Chairman Lee Byung-chul had earlier agreed. The government's economic plan was a systematic way to oversee and assist – that the promises of the 'accused businessmen' were carried out to develop an economically strong and prosperous Korea.

In the government plan, President Park's economic priorities were: (1) develop energy industries, (2) expand agricultural production and increase farm income, (3) develop basic industries, (4) develop export-oriented businesses, (5) improve technology. (Kim et al, 2011) To plan and govern this, Park had established his Economic Planning Board (EPB). Lee Chong Sik wrote: "Firmly in control of political power, President Park recruited professionals to manage the economy. Park's most important decision was to create the EPB . . . (which) controlled both government budget and planning, becoming the central clearinghouse for all economic policy." (Lee, 2012)

President Park did not appoint politicians (nor fellow soldiers) to run the EPB. Instead he appointed professional personnel, technically qualified to formulate economic policy. Among these was O Won Chol. O Won Chul was born in Hwanghae Province and attended the Kyungsung Techology College (now the College of Engineering at Seoul National University). Graduating in 1961, he was hired by the Supreme Policy Council and appointed director of the Chemistry Department. In 1964, he became Director-General of the Industry Bureau and led Park's policy to reorganize Korea's industrial structure toward export. Next he was appointed Director of the Planning and Management Division within the Ministry of Commerce and Industry. Later in 1971, Park appointed him as the Second Senior Presidential Secretary for the Economy. (O Won Chul, 2009)

In 1954, U.S. aid to South Korea had been $82 million and rose to $323 million in 1957, but then declined to $154 million in 1961. The aid had not stimulated economic growth in South Korea. O Won Chul described this: "The Korean economy was at that time marred in a profound slumber. In 1960, the Korean government exported $33 million dollars worth of goods while importing $344 dollars worth . . . importing approximately 10 times more than exporting." (O Won Chul, 2009)

Earlier after the 1962 coup, when the ruling Revolutionary Government established a Supreme Council for National Reconstruction, it had approved a 5-year economic development plan. But the plan didn't work out. O Won Chul wrote: "The year 1963 was a gloomy in Korea. While the government continued to promote the 1st 5-year Economic Development Plan, it soon found itself faced with the worst possible scenario in the form of a shortage of the funds needed to implement the plan. . . . (and) the government was incapable of introducing any international loans to help alleviate the situation. (O Won Chul, 2009)

In December 1964, President Park visited West Germany, having received an invitation from Heinrich Lubke, President of the Federal Government of Germany. Park wanted to see what was called then 'the Miracle on the Rhine', West Germany's amazing economic recovery in nineteen years from the devastation of World War II. Park came away convinced that certain values were important to a nation's economic development. O Won Chul wrote: "the German people's spirit of thriftiness and diligence . . The Germans did not just talk . . . they truly embraced thriftiness and frugality as their basic way of life and regarded extravagance and waste as crimes." (O Won Chul, 2009) And the West German economy was exporting its products to the world.

After the visit, President Park was even more convinced the South Korean economy had to become focused toward an export economy – drawing upon the thriftiness and diligence of the Korean people. In 1964 he had appointed O Won Chul as Director-general of the Industry Bureau, with the task of transforming industry to an export-orientation. O Won Chul saw the government role as: (1) encouraging 'export-oriented' management style in private enterprises, (2) government regulations on licenses and taxes to encourage export, (3) government subsidization of interest rates to facilitate investments in the production of export-goods, (4) government technical assistance to industries desiring to build production facilities for export-goods, (5) economic targets to monitor the progress toward an export economy. (O Won Chul, 2009)

In this list, we see that government was not going to build state-owned enterprises but was to facilitate private entrepreneurship toward export industries. This was Park's principle of a government-industrial economic partnership. There was to be government planning of a desirable national economic policy, but the private capital implementation of economic activity (capitalism focused upon national growth). In 1966, the Park government first five-year plan was completed and was revised in 1967, toward an export target of $350 million dollars.

In 1965, to support the U.S. war in Indochina, President Park sent a marine brigade to South Vietnam and, in 1966, an army division. Park Kisung wrote: "The United States had experienced the risk of communism spreading through the Korean War and was concerned about . . . Vietnam. President Johnson hoped South Korea would contribute to prevent communist expansion in Vietnam. In 1965, he sent his personal letter of 'encouragement' to Korea and expressed his desire that Korea dispatch a combat division to Vietnam." (Park, 2008)

The United States promised to help South Korea with (1) modernization of Korean armed forces, (2) new equipment to counter North Korean infiltration, (3) assumption of expenses for Korean military participation in Vietnam, (4) a loan fund from the U.S. to Korea, and (5) employment opportunities in Vietnam for Koreans. Over 320,000 South Korean troops aided the U.S. troops in the war. The United States government paid the salaries of the South Korean military personnel, and some Korean chaebols received contracts for construction of U.S. military facilities in South Vietnam. These contracts continued to help develop Korean construction experience at international standards and in operating in foreign countries. Thus Korean companies operated in international conditions, as a preparation for the coming global economy.

While this was happening, the conflict with North Korea continued. Kim Il-sung had announced his seven year economic plan in 1961, along with substantial aid and assistance from the Soviet Union. South Korea and North Korea were in a deadly race toward economic growth – capitalism versus communism.

In 1967, North Korea established a Special Guerrilla Corps to train groups of saboteurs to attack South Korean economic facilities. On January 21, 1968, thirty-one commando troops of this Special Guerilla Corp attacked the South Korean presidential residence (the Blue

House) to assassinate President Park. Earlier that month on January 16th, the group had infiltrated the DMZ boundary at the 38th parallel and set up camp on Simbong Moutain, outside Seoul. On January 20th, the troops secretly entered Seoul and assembled in Seungga-sa Temple. The next day, at 8 pm, they approached a police checkpoint near the Blue House. The police chief grew suspicious of the group. He drew his pistol and was shot by the North Korean commandos. They continued to fire at the police and throw grenades at the checkpoint. Thwarted, the commandos ended their attack and escaped. On the next day, the South Korean Army searched for the commandos, finding them hiding on nearby mountains, in total killing 25 out of the original thirty North Korean infiltrators. (Later in 1971, a South Korean commando group 'Unit 684' was set up to try to assassinate Kim Il Sung but it was disbanded.)

On August 15, 1974 on a national holiday celebrating Korea's independence from Japan, President Park and his wife Yuk Young-soo were on the stage in the National Theater in Seoul. An assassin, Mun Se-gwang, had entered the theater, intending to shoot Park. While Park was speaking, Mun moved down an aisle to get closer to the stage on which the president was speaking. Accidently, the assassin fired his revolver, injuring himself. Wounded, he hobbled further down the aisle and began firing wildly at the President. One bullet struck Park's wife, killing her. The assassin Mun was captured and held for trial. He had been born in Japan and lived in Osaka. He had been recruited and trained by an official of North Korea to assassinate the South Korean president. He was hanged.

Thus the war between North and South Korea continued. It was in this kind of hostile relationship, that President Park planned to economically develop his country, not only for his people's welfare but also for their security from communist North Korea.

KOREAN ECONOMIC GROWTH AND MIDDLE EAST OIL

The government policy to create Korean economic growth did not go smoothly. After two five year plans of the Park government from 1962 to 1972, the South Korean economy had improved exports, from $120 million dollars in 1964 to $1 billion dollars in 1970. But the growth had been principally in 'light industry' (such as textiles, shoes, etc.). Then in 1972, Park decided to restructure Korean industry toward 'heavy industry' (such as steel, chemicals, machinery, automobiles, etc.). There were two reasons for this policy: (1) to reduce the need for foreign exchange to purchase imported heavy industry products, and (2) to build a military weapons industry, independent from foreign sources of weapons. To develop the manufactured parts for an effective weapons production capability, South Korea would need to develop heavy industry. We next review the Korean events from 1972 to 1979, as indicated in Figure 5.2.

On November 10, 1971, President Park elevated O Won Chul to be the 2nd Presidential Secretary for the Economy and asked him: "Mr. O! What industry should we develop to reach a 10 billion dollars export mark?" (O Won Chul, 2009)

Mr. O had been preparing himself for this question: "I decided that it was time to present my report on the strategy to advance the industrial structure, a report which I had long been preparing. . . I said: 'Mr. President! I think the time has come to develop the heavy & chemical industry. The Japanese government concentrated on export oriented industries, such as he light industry, as the first stage of its plan to rebuild its economy after World War II. . . . Once it had achieved the 2 billion dollars in export mark, Japan transformed its industrial structure and focused on a heavy & chemical industrialization policy. This happened in 1957. Ten years later, Japan was able to export 10 billion dollars worth of goods. Korea's present situation is in many ways similar.'" (O Won Chul, 2009)

Thus began the transformation of Korean industry from light to heavy industry. But there was a problem. President Park's constitutionally allowed two five year terms, ended in 1972. He would not be in power to carry out a Korean government plan to foster a chemical and heavy industry. He decided to change the constitution – his Yushin constitution.

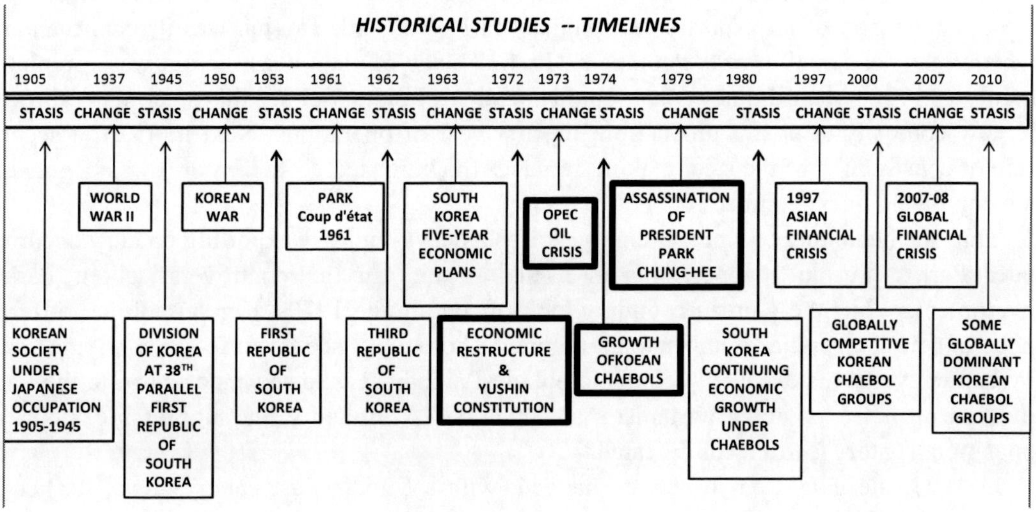

Figure 5.2. Change Events in Republic of South Korea.

In the first term of Park's presidency in 1963, the Republic of Korea passed a constitution which allowed only two terms in presidential office. In 1971, the National Assembly (dominated by the Democratic Republican Party) amended the constitution to permit a third term for Park. Then in the 1971 presidential election, Park won narrowly over an opponent, Kim Dae-jung.

Next in October 1972, President Park now in his third term as president, declared a state of national emergency. He suspended the constitution and dissolved the National Assembly. Park had a new constitution written, which he called the Yushin Constitution. Park chose the term 'yushin' to mean 'renewal' (and it is also the Korean term used to translate the Japanese term for the Meiji Restoration). It was passed in a referendum by 91.5% of the voters (likely in a controlled and not free election). Under the Yushin Constitution, the people elected delegates to an electoral college, which elected the president. The president had the power to appoint one-third of the National Assembly. Park was now elected for a term of six years, with no limit to re-election. He had sweeping powers to suspend constitutional freedoms at his discretion. Ironically, Park followed the same political path of the predecessor, whom he had so despised.

In 1971, Kim Dae-jung (Park's earlier presidential opponent) fled to Japan, where he organized political opposition to the Park's 'Yushin' government. In 1973, Korea Central Intelligent Agents (KCIA) kidnapped him from Tokyo hotel and flew him to Korea. Japanese officials learned of this and protested to the Korean government. Kim was released; and he lived in Seoul until 1976, when he was arrested and sentenced to five years in prison. Later in 1978, his sentence was reduced to house arrest. Political opponents to the Yushin government were harshly treated by the KCIA.

In 1972, President Park had remained in power and continued his plan to develop a Korean heavy industry. O Won Chul wrote: "The blueprints for the construction of the heavy & chemical industry were finally drawn up by the end of the year (1972). These included: plans to develop the shipbuilding, machinery . . electronics . . . steel and chemical industries." (O Won Chul, 2009)

We recall that POSCO was the private company funded by Park's government to build the heavy industry of an export steel company in Korea. Park Tae-jun was its chairman and managed the construction of the company. Its first stage of plant construction was completed on July 4, 1972, when the new plant could finish steel plate. Selling this steel plate allowed the new company to post a profit from its first year of operations. Next in 1973, a second stage of construction of the steel making furnaces (making steel from iron ore) was begun and then completed on November 30,1978.

Thus the beginning of Korea's emergence as a heavy-industry exporting country occurred under Park's Yushin regime. This is a major point in the controversy about Park's government. While the Korean economy took off during the 1970s, Korean politics suffered under a non-democratic, authoritarian regime. Those who see historic good in the Park government value his excellent economic policies; and those who see historic bad in the Park government hate its authoritarianism and the KCIA brutality in suppressing political opposition. History is often complicated.

In 1973, the global environment changed, with a Middle East energy crisis. Earlier in 1960, many nations exporting oil joined in an oil cartel, called the Organization of the Petroleum Exporting Countries (OPEC), in order to control international oil prices. These included: Iraq, Kuwait, Iran, Saudi Arabia, Venezuela, Libya, United Arab Emirates, Qatar, Indonesia, Algeria, Nigeria, Ecuador, Angola, and Gabon.

Then in 1973 Egypt attacked Israel in what came to be called the 'Yom Kippur War'. Syria, and Jordan joined in the war. At first Israel lost many aircraft and appealed to the United States for additional aircraft, which the US then sent to the besieged Israel. Resupplied with aircraft, the Israel turned battle around and won the war. Angered, the Arab members of OPEC cut exports of oil to the U.S. and Europe.

This sudden decline in international oil supply also impacted South Korea. O Won Chul wrote: "OPEC announced that it would separate oil consuming countries into pro and anti-Arab countries, and only those countries deemed to be pro-Arab would enjoy oil supply levels similar to those that existed prior to September 1973 . . . Korea was classified as an anti-Arab state. . . . Gulf Corporation on November 6th notified the Korean government that the amount of crude oil shipped to Korea would be reduced by 30%. . ." (O Won Chul, 2009)

The OPEC embargo on oil to the United States and Western Europe immediately raised international oil prices from $3 dollars a barrel to $12 dollars. And gasoline rationing was introduced in the United States in October 1973. For Korea, the 30% reduction in oil supply would stop its economic growth and inflate its prices. O Won Chul wrote: "Suddenly, I was urgently summoned by Chief Secretary Kim Jeong-yeom who . . . took me to the President's study. Upon entering the study, I saw President Park sitting at his desk . . . Kim neared the President and said 'Mr. President, I think we should dispatch O Won Chol to spearhead our efforts to secure crude oil. His role in the construction of oil refineries has made it possible for him to establish good relations with high-ranking individuals within the three major oil companies.' . . . The President stared quietly for a long time before finally saying 'Just go! You leave tomorrow.'" (O Won Chul, 2009, p. 265)

O Won Chul arranged interviews with the chairpersons of Gulf Oil, Caltex, and Union Oil. At Gulf Oil Corporation, he met with the Chairman, Mr. Dorsey. Mr. Dorsey responded: "I will talk to the person in charge of oil supplies and get him to draw up a plane to help increase the flow of oil to Korea." (O Won-chul, 2009, p. 273) O Won Chul also talked with the heads of Caltex and Union Oil. O Won Chul returned with the good news to Korea and reported to President Park, with the proviso that South Korea must then establish good relations with the Arab States. President Park sent an envoy to the Middle East to begin establishing good relations.

Still oil prices had increased and with this, prices of everything else in Korea. Korea still had to pay higher prices, and the next problem was earning more foreign currency, more exports, to pay for all this. O Won Chul thought that with the increased wealth in the Middle East, there would be increased export opportunities: "The increase in the price of crude oil had resulted in filling the coffers of the oil producing states in the Middle East with dollars. The word on the street was that these Middle Eastern countries planned on using this windfall to develop their economies. . . . Japan was taking steps to gain access to the Middle East in order to reap . . . economic benefits. . . . I knew this was the answer to our problems . . Why couldn't we (Koreans) get in there ahead of other countries and secure an (economic) bridgehead for ourselves? I set about establishing . . . a plan." (O Won Chul, 2009)

President Park invited the Saudi Minister of Planning and Development to Seoul to discuss a future trade agreement in February 1974. Following this, a Korea-Saudi-Arabia Economic Cooperation Council was established between the two governments. In April 1974, a Korea delegation visited Saudi Arabia, and the delegation included business representatives from seven Korea chaebols. To finance Korean companies' bids for Arabian construction contracts, the Korean government would guarantee the construction bonds. This was another example of the partnership between the Korean government and the Korean chaebols which would help build a globally competitive Korean industry.

The first contracts were for road construction in Arabia, and Samhwan Construction won the first Korean contract. O Won Chul tells this story: "One day about a month after the project had begun in September 1974, the Moy of Jeddah requested the project be completed by December 29th when the annual Hajj was set to begin. Samhwan responded to this demand by beginning to work at night using torch lights. The members of the construction crew worked through the night. The use of numerous torches had the effect of creating scenic landscape. . . . One day, King Faisal happened to come across this procession (of torches at night). . . he ordered that more projects be given to these people who worked so hard and with such earnestness." (O Won-chul, 2009)

Shinwon Construction won the first contract to build a harbor in the Middle East in Iran. In October 1975, Hyundai won a harbor contract in Arabia to build an Arab Shipping and Repairing Yard. This harbor project was valued at $137 million dollars and involved 900,000 Korean workers. Hyundai next won a contract to build the Saudi Naval Offshore facilities and next a port in Jubal.

Korean companies participated in the many different kinds of infrastructure projects in the Middle East. In addition to road and port construction, other projects included plant construction, telecommunications, architecture, water supply, energy. These projects helped build the engineering capability of Korean industry, such as Hyundai Engineering and Construction, Daerim Engineering, Lucky Engineering. The value of Middle East projects to Korea rose from $89 million in 1974 to $12.7 billion in 1981.

The economic success of Korea was not that of an 'isolated' nation, but of a nation actively and aggressively participating in global events – gaining economic power by competing in a global opportunities and markets.

The Korean government policies during the 1970s were to build private Korean industry and capability. O Won Chul summarized the development steps of Park's policies: "1) construction of the defense industry, (2) construction of heavy and chemical industry, (3) reorganization of technical education and establishment of research institutes, (5) development of plant engineering industry." (O Won Chul, 2009)

By 1978, Park's economic policies were achieving economic success, but his political policies were not achieving political success. Lee Chong Sik wrote: "The December 1978 general election for the National Assembly revealed Park's government to be a fragile system held in place by sheer force. . . . (the) Yushin constitution . . . was unpopular, as (it had) . . . made Park a perpetual dictator with the power to appoint one-third . . . of the National Assembly. These measures . . . heightened the discontentment among the populace, including labor, students, and intellectuals. The government harshly suppressed all forms of opposition, but it was clearly fighting a losing battle. The opposition's New Democratic Party won a handsome plurality with 34.7 percent of the vote. . . The majority of the voters clearly wanted a change." (Lee, 2012)

Park's opponent, Kim Young-sam, attacked Park's government, and the government retaliated by removing Kim from his seat in the National Assembly. This further intensified the opposition's struggle in the streets. Lee wrote: "Applying more pressure to the opposition would create martyrs . . . loosening control might foster the spread of demands for reform . . Park's lieutenants were divided on which measures would best manage the crisis. While they were deliberating these measures, the director of the KCIA, Kim Jae-kyu, shot President Park . . . dead on December 12, 1979." (Lee, 2012)

PERCEPTUAL SPACE OF THE EVENTS OF PARK'S GOVERNMENT

Now we can summarize the principle events of President's Park's governance in a societal dynamics perceptual space, Figure 5.3.

INDIVIDUAL – The principle individual in these events is Park Chung-hee, a Korean military officer who seized government in a coup, temporarily re-established democracy and was elected for two terms; but he then altered the constitution (Yushin) in order to be re-elected perpetually. He was assassinated by the head of the Korean Central Intelligence Agency.

SOCIETY – The societies involved included the Korean society, divided into the two nations of North and South Korea by the ideologies of communist dictatorship against capitalist democracy.

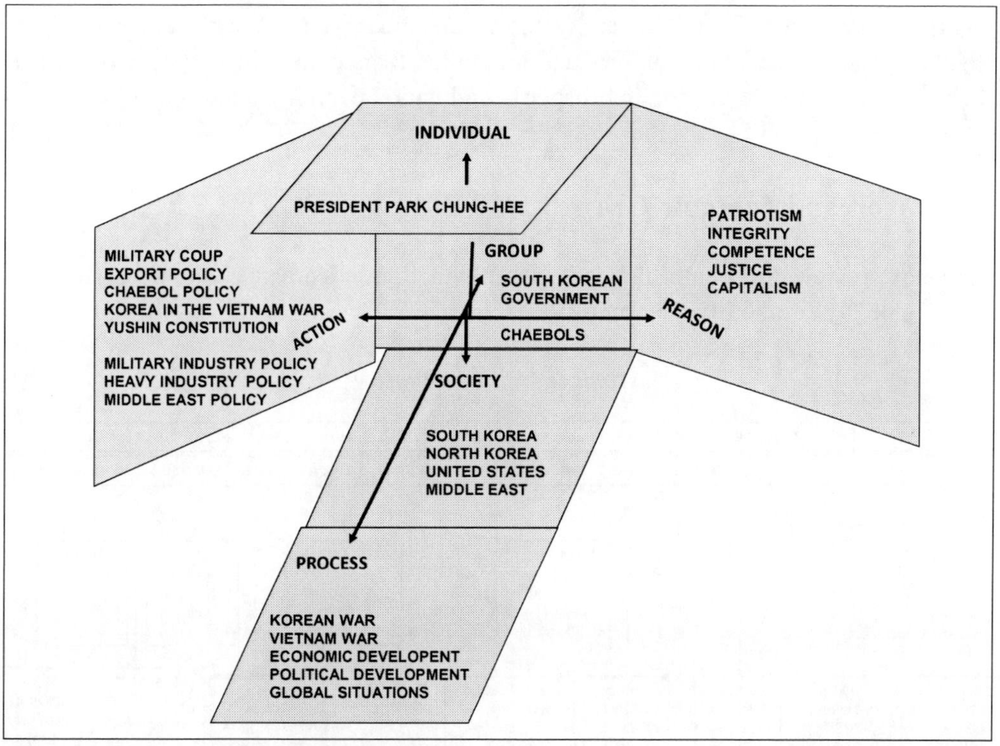

Figure 5.3. Perceptual Space of Events in Park's Government.

GROUPS – In South Korea, the principle groups involved in the events were the South Korean government led by President Park and the industrial groups of the Korean chaebols, family-controlled groups of enterprises).

PROCESS – The processes involved in the events included: (1) the Korean War, establishing enduring military and political conflicts between the northern and southern governments of Korea, (2) South Korean participation in the Vietnam War, (3) economic development of South Korea, (4) political conflicts in South Korea between democratic governance and authoritarian governance, (5) global opportunities of the Vietnam War and Middle East infrastructure construction.

ACTION – The actions involved in the sequence of events under Park's governance include: (1) military coup to seize governance of South Korea, (2) policy to develop Korea as an export economy, (3) policy to focus Korean chaebols toward export industries, (4) Korean support of the United States in the Vietnam War, (5) Yushin constitution to enable Park's continuing presidential reelection, (6) policy to develop Korean weapon industry for military independence of South Korea, (7) policy to develop Korean heavy industry, and (7) policy for support of Arabs versus Israel, in order to gain access to construction projects in the Middle East.

REASON – The reasoning underlying Park's policies in governing South Korea from 1962 to 1978 include: (1) Park's patriotism as a citizen and soldier of South Korea, (2) Park's insistence on integrity in government officials and in the economic actions of chaebol leaders, (3) Park's appreciation of competence by government professionals (such as O Won-chul), (4) Park's desire for social justice for all Korean people, business, labor, farmers and dislike of the older 'yangban' feudal hierarchy in Korean history, (5) Park's understanding that capitalism (private ownership of production) rather than communism (state ownership of production) would lead to economic prosperity and national strength and social justice.

History: Korean Economic Growth 1980-2013

We next review the continuing growth of the South Korean economy after 1980, to become a globally competitive nation by 2013, Figure 5.4.

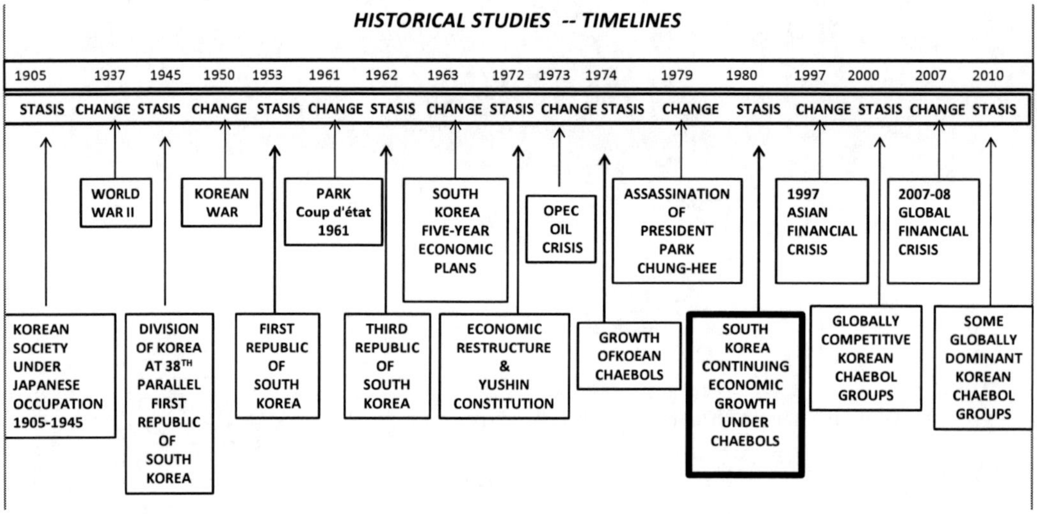

Figure 5.4. Change Events in Republic of South Korea.

The chaebols continued to grow and develop export industries in accordance with government policies throughout the 1970s and 1980s, even after the death of Park. One factor in their growth was favorable financial conditions which the government provided to Korean businesses for encouraging manufactured exports and construction contracts in foreign countries. Kang Myung Hun wrote: "During the 1960s and 1970s, the Korean government exercised substantial control over all formal sources of credit, both domestic and foreign. Government industrial policy dictated the credit decisions of Korea's major city banks, which had been nationalized in 1961 as part of the government's campaign against illicit wealth accumulation' during the former regime. . . . Government intervention in the allocation of credit increased considerably in the 1970s, subsequent to the major reductions in interest rates mandated by the August 3 Presidential Emergency Decree of 1972. . . . In general, the system of credit rationing in the 1970s favored large firms at the expense of small and medium-sized firms. . . . by controlling the allocation of credits, the government could actively support the growth sectors." (Kang, 1996)

From 1962 to 1989, South Korea's gross domestic product grew from \$2.7 billion to \$230 billion, a multiple of 100 times. Posco founded in 1968, sold its first steel plate in 1972 and in 2010 had grown to be the fourth largest steel industry in the world. Samsung began its R&D to produce semiconductor memory chips in 1983 and by 2010 would dominate the world memory chip market. Hyundai founded again in 1946 became by 2010 the world's largest ship-building company and the fifth largest automobile producer.

After Park Chung-hee's assassination, General Chun Doo-hwan, the chief of the Defense Security Command, arrested the assassin, tried and executed him. Chun Doo-hwan, with a group of other generals, then seized the government in a second military coup. Chun Doo-hwan declared martial law and arrested political opponents, Kim Dae-jung and Kim Young-sam. In 1980, Chun was elected president by the national assembly, under the Park's Yushin constitution. Chun served a five year term as president to 1986. But the Yushin Constitution was still in the law; and in June 1987 a new protest movement of students and citizens, June Democracy Movement, wished the restoration of direct presidential elections.

Roh Tae-woo, another former general, was the government's choice to follow Chun as president. On June 29, 1987 as the protests had continued, Roh announced a Declaration of Political Reforms and authorized again a direct presidential election. Roh was elected president in the election of 1988. Roh Tae-woo had been another military general, serving from 1955 to 1981, and had been an officer in Korea's 9th Division which served in the Vietnam war. Roh served one term as president, from 1988 to 1993.

In the first direct presidential election in 1987, two opposition leaders, Kim Young-sam and Kim Dae-jung had run against each other and split the opposition vote – so that former general Roh Tae-won had become president. In the next election in 1990, Kim Young-sam merged his party, Peaceful Democracy Party with Roh's ruling party, Democratic Justice Party, to win the election, defeating Kim Dae-jung.

But Kim Young-sam turned upon the Roh Democratic Justice Party and launched an anti-corruption campaign – requiring government and military officials to publish their financial records. President Kim Young-sam had both Chun and Roh arrested on charges of corruption. In August 1996, they were convicted. Chun was sentenced to death but commuted to life imprisonment; and Roh received a 22 year sentence, reduced to 17 years. A year later in December 1997, both Chun and Roh were released from prison, pardoned by President Kim Young-sam In 1997, the Asian Financial Crisis hit Korea in the last year of the term of Kim Young-sam. Kim Dae-jung became the next president.

So by 1997, Korean economic growth depended fundamentally upon chaebols, acting in a strategic partnership with government. Kang Myung Hun wrote: "The structural domination by big firms (in Korea) is manifest in the serious imbalance between small and medium-sized firms and big firms. The growth rate of the output of small and medium-sized firms was significantly lower than the average growth rate for the economy throughout the 1960s and 1970s." (Kang, 1996)

Also the business growth of the chaebols had funded growth with large amounts of loans. Kang added: "Korean firms in general are highly leveraged the debt-equity ratio of the average manufacturing firm is 4 to 1. The larger a firm is, the higher is this ratio because of credit-financed expansion. . . . If a large Korean enterprise gets into trouble, the whole group, which is so large that it cannot be allowed to fail, is endangered. Therefore, the existence of large groups forces the government to implicitly or explicitly guaranteed group credit." (Kang, 1996)

When in 1997, the global Asian financial crisis occurred, South Korea was caught in the panic, and the Korean government intervened to force Korean chaebols to 'clean up' their debt situations. (We review this in a later chapter.)

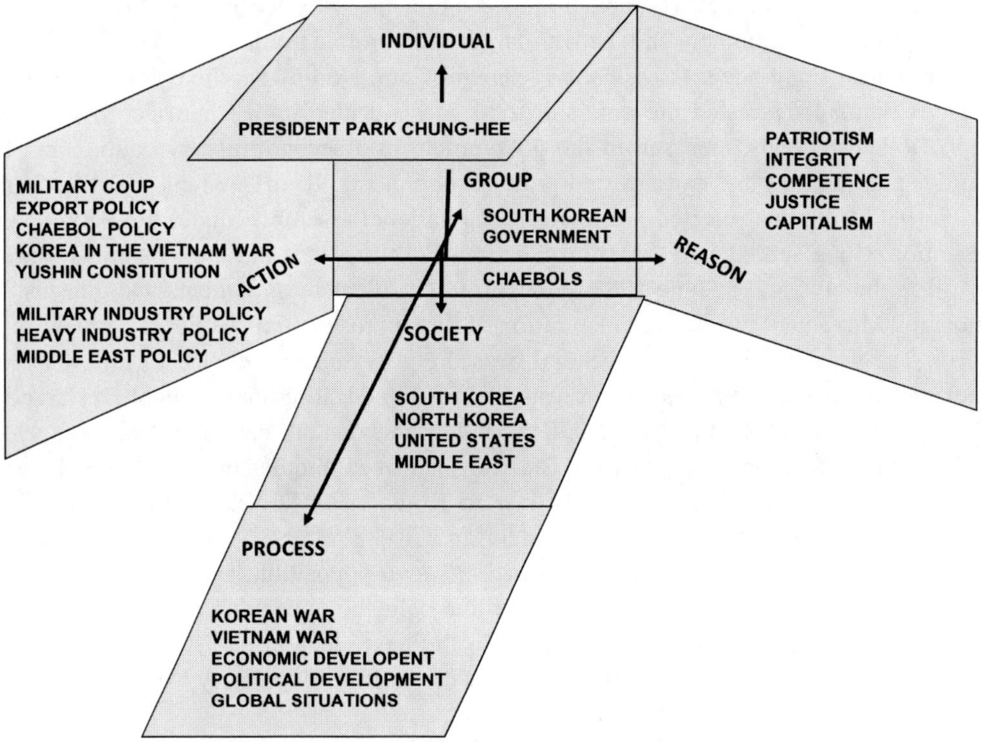

Figure 5.5. Perceptual Space of Events from 1978 – 1997.

Control in Korean Government Policies for Economic Self-Organization

Under Park's government, South Korea was building a self-organizing economic system. The export market (free-market) determined economic winners and losers among Korean chaebols. The government role in market organized economic system is one of regulation and monetary policies. To analyze the kinds of control which operate in self-organizing societies, we use societal dynamics theory. (Betz, 2011) The theory is summarized in the topological graphs, of the earlier Figure 5.5 and also in Figure 5.6.

This framework enables the analysis of any historical event of a society in terms of the kinds of control which occur in a self-organizing society, such as a capitalistic democracy. Businesses or government agencies exercise direct control over their organizations (controlled as managed systems) of each business or government agency. The socio-technical systems of a society, such as the economic system, operates as a total result of all the economic actions of all the businesses, each acting as an individual economic agent. This total result of economic actions is bounded by the regulatory and infrastructures of the society, which are in turn impacted by government policies in regulation and infrastructure development. The bound of the self-organizing aspects of society (bottom-up perspective from the Society end) can be described as the 'control in the self-organizing system. It is this self-organizing control which Park's government policies constructed in the economic development of South Korea.

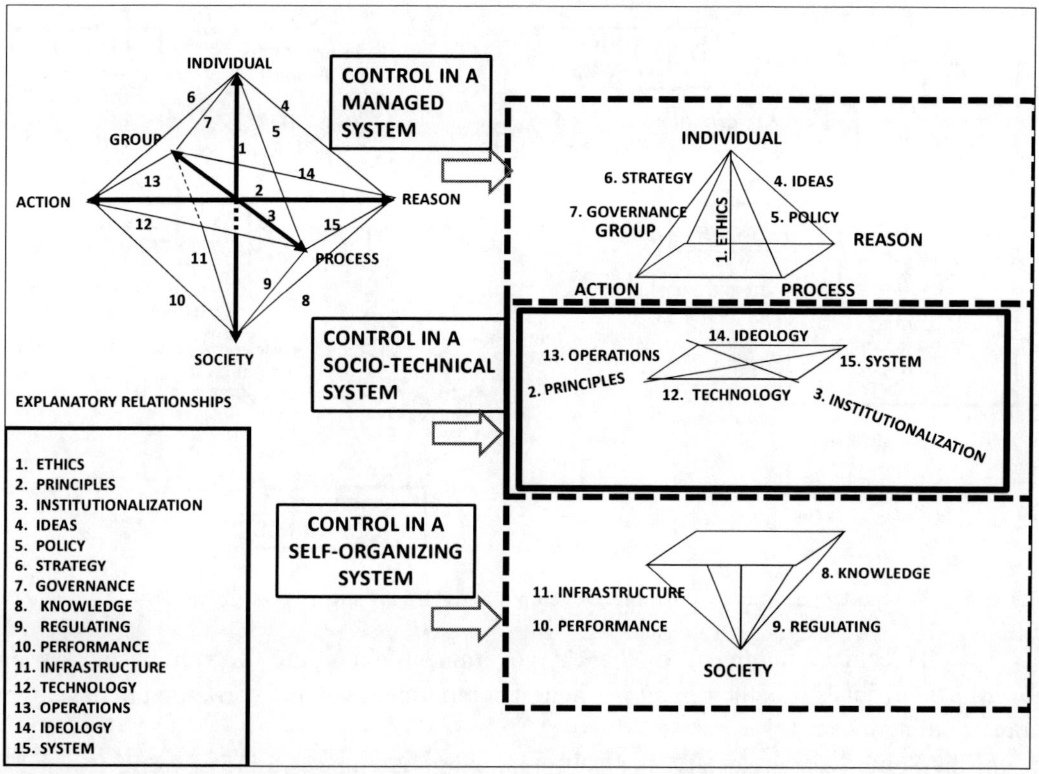

Figure 5.6. Types of Control in A Society.

We use this analytical framework to understand President Park's policies for the development of an export-oriented manufacturing economy in South Korea, shown in Figure 5.7. Therein the government role in society can be explained by analytical concepts of a self-organizing system of capitalism encouraged by Park's government and of the socio-technical system of the South Korean economy. Governments rule the self-organizing forms of society, intended to implement the functional needs of the society, as national socio-technical systems.

General Park had seized the South Korean government in a military coup, establishing himself as president in two directly-elected terms and two indirectly-elected terms. President Park's policies were to develop a strong South Korean economy in manufacturing for exporting goods and services to a global market. Park's policies encouraged the development of an business infrastructure of capitalist chaebol, as long as each had development plans in correspondence with the government's plan to build a strong exporting economy.

The self-organizing system was the society of South Korea. In that self-organizing system, the key explanatory relations are *knowledge, regulating, ethics, infrastructure and performance*. We can see from the historical narrative of Park's government that (8) Knowledge for economic development required Korean firms to obtain technology from other nations, *foreign technology*. The government's (9) Regulating was expressed by a national economic development plan, *EPB national plans*, as a guide to which industries Park's government would assist private enterprise to develop. In that plan, the EPB focused upon developing the (11) Infrastructure of Korean businesses in *heavy industry*. The overall (10) Performance of Korean industry was measured by the Park government as the quantity of *exports*.

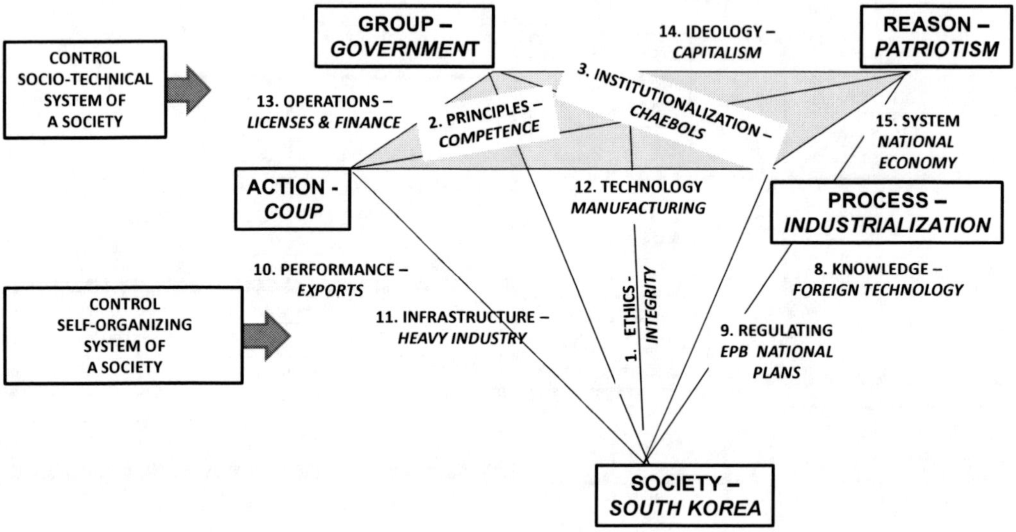

Figure 5.7. Korean Government Control to Develop a Self-Organizing Export Economy.

The (1) Ethical judgment by Park's government about the Korean businesses and government officials was their *integrity*, honest commitment without corruption to developing Korea's strength.

In the South Korean society, this self-organizing system was a (15) System focused by the Korean government as its *national economy*. The (3) Institutionalization of the economic socio-technical system was through *chaebols. T*he (2) Principles used to judge these were their economic competence. The (12) Technology of the Korean socio-technical economic system focused upon *manufacturing*. The (14) Ideology of the South Korean socio-technical economic system was *capitalism*. The (11) Operations used by the government to steer chaebol focus upon goals of the national economic plan were *government licenses and access to credit* – government licenses permitted a company to enter a business and to export its products. Government assistance with credit helped a business to finance that development.

Control in Managed System of Chaebols

Next we analyze how the chaebols were controlled by their founders, as a managed system in a business organization, Figure 5.8. In the authority structure in the managed system of a Korean business organization, the Individual in authority is the chaebol leader (founder or CEO). The chaebol is a Group of interlocking companies controlled by one chaebol family; and the Reason for the group is the accumulation of wealth. The Action of the Group is as businesses enterprises, whose Process is production of goods and services.

In such a Korean business organization authority-structure, the (7) Governance of a chaebol is through *family control*. The (6) Strategy of a chaebol was business growth, for the (5) Policy of producing *export goods and services*. The (4) Ideas within a chaebol are *entrepreneurial*, to start or acquire new businesses; and the (1) Ethics of the chaebol is *integrity* in providing quality products.

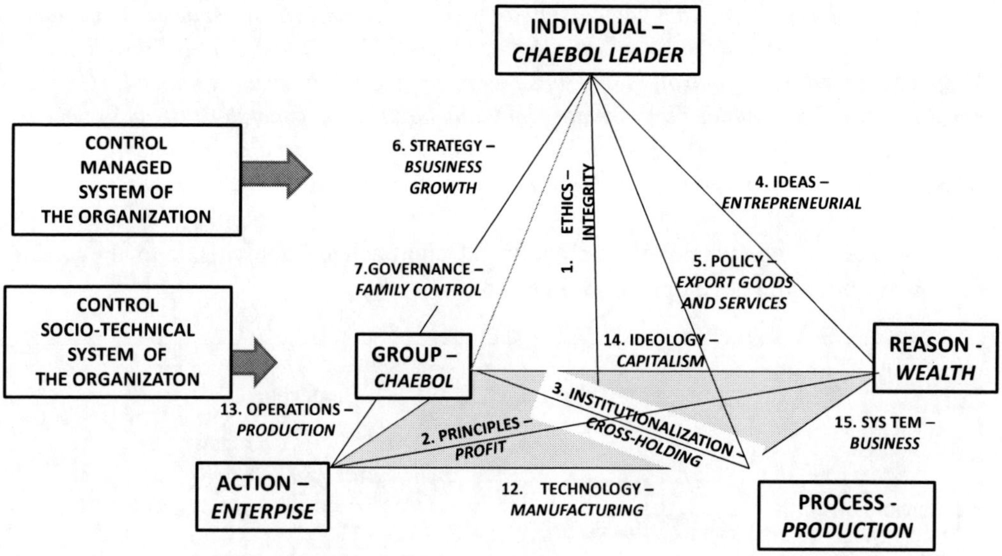

Figure 5.8. Managed System Control in a Chabol Group.

In addition to an authority structure, an organization has operations expressed as the socio-technical system of the organizations' business function as a (15) System of *business.* The (13) Operations of the Korean chaebol businesses under the Park regime were focused upon *production*, particularly in heavy industrial products. In (12) Technology for this production, Korean chaebols would at first acquire foreign *manufacturing* technology in the early period of Park's regime but turned next to developing indigenous technical progress through Korean research and development. The (2) Principles of the business system of the chaebol enterprises was to create *profit.* The (3) Institutionalization of a chaebol group was through *cross-holding* in multiple business corporations.

SOCIETAL CONTROL IN THE SOUTH KOREAN ECONOMY

Thus the challenge of societal control can be divided into three control issues: (1) control of managed systems (i.e. organizations), (2) control of socio-technical systems (i.e. economic systems), and (3) control of self-organizing systems (i.e. a society). The advantage of looking at control in a society in these three different ways is to separate issues of organizational control (businesses as managed systems) from regulatory control (government regulation of an economy) from functional effectiveness in a society (economics as a socio-technical functional system of a national economy). We have seen the contrast between 'control' in a private business (direct control) and societal 'control' by a government agency (indirect control). In this historical case, we also have seen the importance of the kind of partnership which had been formed between the South Korean government and South Korean industry – a partnership which President Park and the chaebols leaders (such as Lee Byung-chull) had together made for the good of the country.

As societal control, the partnership between government and industry was this. Government controlled the 'self-organizing society' of South Korea, through (9) Regulating. Industry controlled the businesses in the society as 'managed systems' through (7) Governance by private Individuals. This partnership of government regulation in a self-organizing capitalism did enable the socio-technical economic system of South Korea to prosper as (10) Performance of an export economy.

We can see the usefulness of this analytical approach by applying it to the period of South Korean economic development, Figure 5.9.

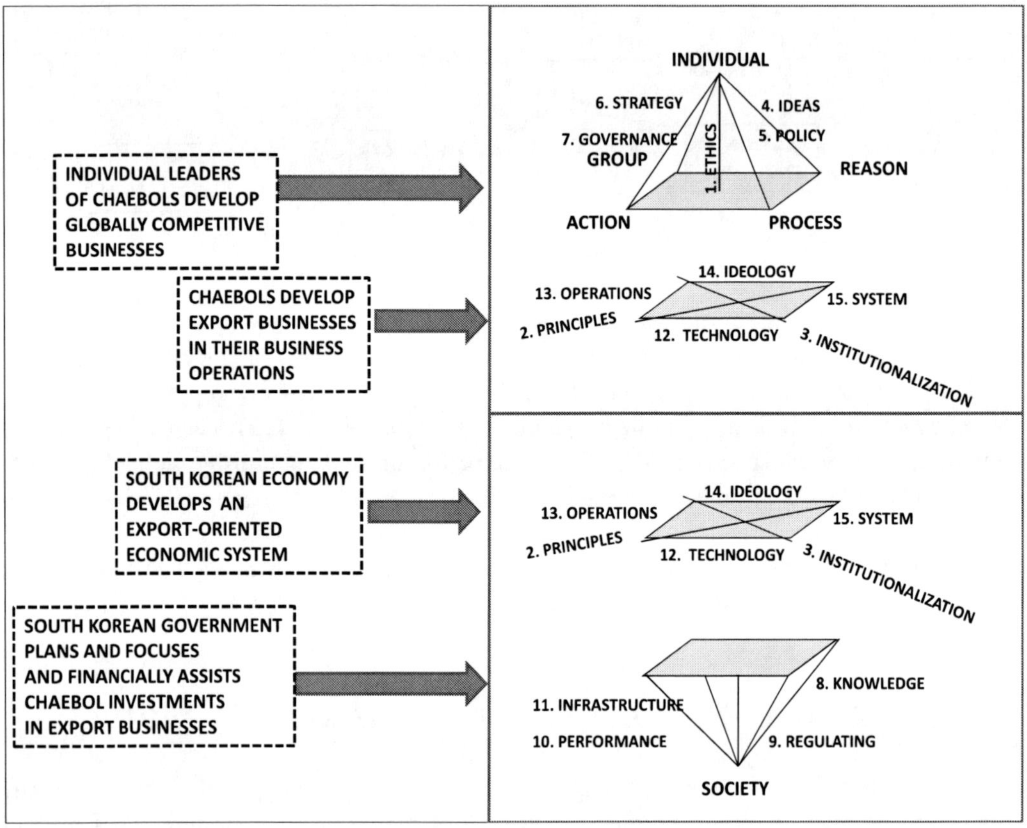

Figure 5.9. Societal Controls in South Korea Economic Growth 1962-1996.

Control in the managed systems of the Korean chaebol businesses had leadership in the authority structure of their organizations with strategies of business growth and policies of export goods and services. Complementary to the chaebol groups, control in the Korean government used national economic plans for regulating the development of the industrial structure of heavy industry with a performance of growth in exports. Thus the controls in the Korean society were divided between (1) the direct controls of the chaebol businesses and government indirect control through regulating the economy. These two forms of control (managed systems and self-organizing systems) were coordinated through a patriotic partnership between the chaebol leaders and the government leaders.

In a society, 'control' must be used to order functional processes, social processes. First within a society, the concept of societal 'control' means to manage a business, an enterprise, an organization, a government agency – managed systems. Second within a society, the concept of societal 'control' means to operate the connections between institutions delivering a societal function (e.g. manufacture of products, retail services, transportation services, energy systems, medical systems, educational systems, public safety, military security, etc.) – socio-technical systems. Third within a society, the concept of societal 'control' means to regulate the functioning of the socio-technical systems of society – self-organizing systems.

> *This is generally the way of control in any capitalist democratic society, direct and indirect – direct in businesses and indirect in government policies. When the two forms of societal control are in coordination, the society can prosper. But when in conflict, society can suffer.*

LESSONS LEARNED

In a developing country, economic growth can be facilitated by an appropriate partnership of government and private industry. The partnership is one in which government facilitates national economic growth by encouraging an export oriented industry, providing the infrastructure for national education and creation of technical knowledge, regulating an honest banking system, not tolerating corrupt officials.

In turn, industry should invest in the production of exportable safe, quality products of advanced technology, provide employment opportunities to the people of the nation with reasonable levels of salary, manage the business with reasonable ratios of debt to equity, and fairly share profits with general shareholders.

Chapter 6

DESIGN IN GLOBAL MANUFACTURING – HYUNDAI

INTRODUCTION

We next examine the extraordinary business success of the Hyundai Group in construction, ship-building, and automobiles. The founder, Chung Ju-young, displayed an extraordinary entrepreneurial style in building the chaebol. And he always operated in a global environment, for opportunities, finance, and technology. His businesses profited by delivering a quality product quickly but at a lower price than foreign competition.

HISTORY: FOUNDING OF HYUNDAI

We recall that the founder of Hyundai, Chung Ju-jung was born in 1915 in a village now in North Korea. His was from a peasant family who worked hard to gain a living on marginal land; but he was determined to escape to the city to gain a more prosperous life. And he recall that after three attempts and another year, he succeeded in finding work in Seoul. He had worked first as a construction laborer, then as a handyman in a starch syrup factory, before getting a job delivering rice. He worked hard and helped to manage the store. In 1937, after the owner became ill, Chung bought the business. He moved his family to Seoul to help him in the rice store. But in 1939, the Japanese occupation further outlawed Korean ownership of agricultural businesses, and Chung lost his shop, without compensation. Donald Kirk wrote: "One of the first decisions of the Japanese after seizing power in 1910 had been to enact the notorious 'Company Law' that made it illegal to run a company without Japanese investment and Japanese in management." (Kirk, 1994a)

Also we recall that a year later in 1940, Chung returned to Seoul, and started again. But this time he bought an automobile service garage, the Ado Service Garage. Donald Kirk wrote: "Chung said, 'When the rice business was forbidden, I was looking for something else to do that did not require large capital but would yield high returns. Fixing cars was just the thing – one of the few businesses that a Korean could run while the Japanese sold the cars and the parts. I went back to the (Korean) rice mill owner and borrowed the money to purchase a repair shop. Whoever fixed a broken car sooner could earn as much money as possible . . . We would repair in five days what others might have taken 20 days to do, and we worked through the night. As long as we fixed promptly, we were paid what we had asked.'" (Kirk, 1994a)

Chung's business strategy was to repair cars more quickly than competitors. This kind of 'quick' strategy was to characterize all his later business endeavors. Yet once again in 1943, Japanese officials intervened. He was forced to give his repair business to a Japanese-owned local steel plant. Still Chung had some savings (50,000 won). He returned to Seoul after Japan lost the war. He set up another garage, which he named 'Hyundai Jadongcha Gongupsa' or 'Hyundai Motor Service'.

Next on Jung 25, 1950, North Korea invaded across the 38th parallel, taking Seoul three days later. Chung Ju-young and his younger brother, Chung In-yung, retreated with the South Korean and U.S. army troops to Busan. Chung In-yung found a job translating for the U.S. Army. Ju Yung found work building tent floors for the U.S. troops. Chung Ju-young next won a contract from the Korean transport ministry to build a warehouse (for the influx of goods from the American army). For this, Chung Ju-young founded Hyundai Civil Works Company, and he also founded a transport company, Hyundai Commercial Transportation, for shipping food and other goods around the southeastern region of Busan.

After the Korean War, Chung continued construction work and became a major contractor In 1957, he won a contract from the South Korean government to reconstruct a bridge across the Han River (which had been destroyed in the war). He won a contract from the U.S. Army to rebuild Inchon Harbor, for both military and commercial shipments.

We recall that Syngman Rhee resigned as President of South Korea on April 26, 1960, forced out of office by a student-led revolt (April Revolution). A year later on May 16, 1961, Major-General Park Chung-hee (Director-General of ROK Army Operations) formed a Military Revolutionary Committee and seized power. In 1963, Park won the presidency of South Korea, and his government embarked upon a plan to build an export economy.

President Park's vision for building a strong Korean economy and Chairman Chung's vision for building industry in Korea coincided. Donald Kirk wrote: "Chung found an ally in Park as a man who respected what Hyundai could do for the economy and was willing to give him and Hyundai the encouragement and support he needed. . . The Hyundai Cement company was a turning point for Chung. His company previously had had no fixed manufacturing facility . . . Now, he had his first factory. . . . This story of Hyundai Cement illustrated how Park and Chung relied on foreign help in pursuit of common goals . . . Chung paid for the factory with a $4.25 million loan from the Korean government, which had borrowed the money from the U.S. Agency for International Development (AID). . . . Hyundai did not have to pay interest for two years, and then had ten more years to pay off the loan at 6.76%. . . . The factory justified the investment for the country and the company, when it exported cement to Vietnam as U.S. combat troops arrived in 1965. The real payoff for Hyundai was that its engineers could learn by observing – and were soon ready to build their own cement plants, some for export." (Kirk, 1944a)

In the construction work, cement was required for building concrete structures. With Hyundai building its own cement factory, it could reduce its materials costs in construction projects.

We recall that President Park decided to assist the U.S. in the Vietnam war, supplying a South Korean brigade to fight. In Vietnam, Hyundai also found construction work for the U.S., working as a sub-contractor under a large U.S. conglomerate, RMK-BRJ (which held the U.S. military construction contracts). In 1967 and 1968, the earnings of Hyundai were large, constituting 10% and then 20% of South Korea's foreign earnings. Donald Kirk wrote: "The war was well-timed for Chung's ambitions, but he also had to have impressed he Americans with a drive not ordinarily found among the horde of Korean contractors. . . . From this raw beginning under American tutelage, Chung would go on to much bigger contracts – in his own country and then Southeast Asia. . . Finally he would display his usual incredible talent for being at the right place at the right time, with the right prices and expertise, to cash in on the bonanza in building accompanying the Middle East oil boom." (Kirk, 1994a)

Hyundai's drive to earn from foreign revenue through overseas construction projects fit nicely with President Park's drive for Korean exports. We recall Park had encouraged South Korean firms to contract for construction work in the Middle East. To assist, Park's government offered financial subsidies on exports, exemptions from taxes, rebates on tariffs levied on imports which were then used for exports. The government protected chaebol products from imported goods with high tariffs on imports. The government operated a national bank to help finance export operations. For all this assistance, government was selective on which chaebols to help – those who were successfully competing, like Hyundai.

Lee Kyu Uck described Park's government policies: "Strategic industries emphasized by the government required huge capacities in technology, capital, manpower and organization – hence business groups (chaebol) were favored over other firms. Thus it was natural that the benefits of the tax, finance and trade policies should accrue primarily to the business groups with growth further aided by the growth-first policy of the Korean government." (Lee Kyu Uck, 1986)

With this convergence of goals, competence and exports, Hyundai Chairman Chung and President Park got along quite well. Donald Kirk wrote: "Chung and Park appeared to many observers as brothers in rapport: 'Chung and Park had quite a similar outlook.' . . . While Park decreed that Korean focus on industrial progress, Chung put the policy into action in his own industries. In the year 1969, making a profit from big business in Korea was not all that easy. A number of enterprises, bloated on policy loans . . . had gone bankrupt. Park had Chung and Hyundai to thank for justifying his economic program in the face of the critics. He saw possibilities for elevating Korea from third-world to world-class." (Kirk, 1944a)

We recall that in 1973, the OPEC embargo on oil to the United States and Western Europe had jumped international oil prices from $3 dollars a barrel to $12 dollars. And for Korea, there would be a 30% reduction in oil supply that would stop its economic growth and inflate its prices. President Park's response was to send a delegation of Korean businesses to find construction work in the newly enriched Arab kingdom. Chung Ju Yung had responded with enthusiasm, first gaining a contract to construct a facility in Bahrain. Donald Kirk wrote: "The Bahrain gambit propelled Hyundai onto the Middle East scene as a serious contender. But it was the project for building the harbor at the Saudi port of Jubail in the Persian Gulf that established Hyundai as one of the leading forces for construction in the region." (Kirk, 1944a)

To build factories in the Middle East, Chung founded Hyundai Engineering. In 1999, Hyundai Construction and Hyundai Engineering were merged. Hyundai continued to build facilities in the oil-rich Middle East states. For example in 1992, Libya Hyundai built four electrical power plants. In Singapore, Hyundai built an electric railroad and a container terminal. Hyundai built a marine terminal for a refinery in Malaysia and a port in Indonesia. Major construction projects by Hyundai Engineering and Construction included: Kyeongbu(Seoul-Busan) express way, North Han river, Soyanggang Dam, multipurpose Dam, Seoul Gangnam apartments, Hyundai Ulsan shipyard, Hyundai Motor Ulsan complex, Ulsan Industrial complex, Posco Giant Pohang and Gwangyang(South-west of Korean peninsula) complex, Jubail Industrial harbor construction, Singapore Suntec tower project. Later, the global locations of Hyundai Engineering and Construction (E&C) included: Hyundai E&C U.S.A.; Hyundai E&C U.K.; Hyundai E&C Japan; Hyundai E&C Hong Kong; Hyundai E& C Singapore; Hyundai E&C Vietnam; Hyundai E&C Indonesia; Hyundai E&C Libya; Hyundai E&C Iran; Hyundai E&C Kuwait; Hyundai E&C Saudi Arabia; Hyundai E&C Qatar; Hyundai E&C UAE; Hyundai E&C India. (http://hyundaiengineering.blogspot.kr/ 2010/11/hyundai-conglomerate-groups-of-korea_7719.html, 2010)

Meanwhile in the 1970s, during port construction in the Middle East, Chung also launched a ship-building business. Donald Kirk wrote: "Chung was getting ready to expand into heavy industry about the time that he was moving into overseas construction. . . If he were going to engage in port construction overseas, why not produce the ships to carry the goods to and from the ports? How Hyundai got the money, and then its first contracts became the stuff of industry legend – an example between big government and big business, Korea-style." (Kirk, 1944a)

Ship building was a good export business, since ships were built to sell to foreign shipping companies. Chung built a new domestic port, for the docks on which to build ships. And this plan fit nicely with President Park's continuing desire to build industry in throughout South Korea, particularly in Ulsan. Donald Kirk wrote: "In one of his first steps to modernize the economy, Park in 1962 had singled out Ulsan as 'a special industrial area.' Supremely confident, Chung in 1970 bought land for the shipyard and . . signed up Appledore and Scott Lithgow of Great Britain for sales, engineering and administrative support on highly favorable terms – $1.7 million for the technology and 0.5% as a commission on sales. The real problem now was how to raise $63 million needed to develop a yard with a capacity of (building ships) of 500,000 ton (size). Chung could only drum up $10 million from his own coffers, while his government . . advanced him $10 million. That left $43 million to go." (Kirk, 1994a)

Chung's extraordinary entrepreneurial skill was to envision big projects and then figure out how to raise the financing. Donald Kirk wrote: "Chung needed Appledore's recommendation in order to get the loan. But the Appledore chairman, Charles Longbottom, was not confident that Hyundai could find buyers; still he wrote a letter to Barkclays Bank, which agreed to provide the loan on the condition that Chung prove he had buyers. . . . (Next to help find a buyer) Longbottom introduced Chung to a shipbroker, who suggested he call on Greek ship owner George Livanos. Chung visited Livanos, bearing a photograph of the proposed site for building the ships and a drawing of one 260,000-ton tanker. Livanos agreed to purchase two 259,000-ton crude oil carriers. . . . (Chung) promised to deliver at prices 15% below those of any other manufacturer." (Kirk, 1994a)

Price and quality and time were the three principal factors upon which Chung always competed. He would bid prices below competitors and deliver a quality product and quickly. But even when global economic conditions did not always meet Chung's time schedule, the South Korean industry-government partnership would be there. For example, the Greek ship owner did not finish buying the ships, cancelling the order and forfeiting a 15% deposit. The reason was the sudden decline in demand for shipping oil after the oil prices jumped due to OPEC. Still South Korea needed to ship oil purchased in Saudi Arabia for Korean consumption. Chung organized a new company, Hyundai Merchant Marine Company in 1976 and used the two oil tankers he had just constructed to deliver Arabian crude oil to the South Korean National refinery. Korean law at the time stipulated that only Korean ships were permitted to deliver oil to the refinery. (Kirk, 1994a) Thus Chung did not lose money on his first ships which he built, and he also started another business.

Donald Kirk wrote: "Hyundai Heavy Industries . . came to lead the world in ship-building. Engineers from Britain and the rest of Europe migrated to Ulsan on contracts after their own companies had gone bankrupt. . . . The yards on the Tyne, home in England's northeast for British ship-building, were almost empty, while Hyundai was now capable of working on 46 vessels at once. . . . In a sluggish market, Hyundai foresaw orders for more and more specialized ships. Hyundai established two research institutes, Hyundai Maritime Research and Hyundai Industrial Research, to keep up with trends and perfect new techniques. . . . Since building its first LPG carrier in 1986, it had produced 3 by 1992. The company in 1993 was pouring $367.2 million into facilities and R&D . . . for specialized ships, off-shore structures, welding technology and on-shore plants and bridges." (Kirk, 1994a)

ENTREPRENEURSHIP AND HYUNDAI

In the business literature, there are discussions of different management styles for starting and for growing businesses – entrepreneurship versus professional management. Different management styles are appropriate in strategy implementation during the different stages of a company's life – entrepreneurship during the creation of a new company and professional management for its growth.

In this literature on entrepreneurship, two themes emerged:

1. There is an emotional theme. The entrepreneur is a kind of business hero/heroine; and like all such, they have qualities to be admired – initiative, daring, courage, commitment. These virtues are especially admired in turbulent business conditions, when initiative is required to originate a business or to make a drastic change for survival.
2. Problems of change of leadership occur, after a founder retires or even during a founder's life, when an organization has grown so large and that it requires more rationalization than the dictates of the founder. Then professional managers begin the running of the operation, some successfully and some unsuccessfully.

For the first theme, one can see that the biography of Chung Ju-yung shows a heroic theme – struggling against servitude as a youth under the Japanese occupation of Korea and then building a global corporation in a free South Korea.

Howard Stevenson and David Gumpert compared managerial strategic styles along two dimensions: (1) desire for future change, and (2) perceived ability to create change. They noted that entrepreneurs ask questions, such as: (1) Where is the opportunity? (2) How do I capitalize on it? (3) What resources do I need? (4) How do I gain control over them? (5) What structure is best? (Stevenson and Gumpert, 1985)

These questions characterize Chung Ju-young's management style. He always asked: What is the opportunity? He saw an opportunity for a rice business. When this was taken away, he saw an opportunity for an auto repair business. When this was taken away, he waited until the liberation of Korea and began again his auto repair business. When the North Korean army invaded South Korea, Chung and his brother retreated south. He saw an opportunity for a construction business for the U.S. Army. He continued to see opportunities, for more construction business in South Korea under President Park and for the Americans in Vietnam. With President Park's support, he next saw construction business opportunities in the Middle East and later Southeast Asia. While constructing ports, Chung saw a business opportunity in building ships.

Both visions of opportunity and capability to get capital were essential to the entrepreneurial ability of the founder of Hyundai. Chung Ju-young wished to start and grow businesses, and this continually required a lot of capital. The advantage of the chaebol form was that cash flow from successful businesses which had been started could provide capital to start and grow new businesses. Also the profitable operations of one company could be used as collateral to guarantee loans to build a new business. In 1950, when Chung gained cash from tent-base construction for the U.S. Army, he could purchase trucks and start a transportation business. In 1967, when Chung began Hyundai Motor, Hyundai Construction was generating the cash flow and assets to fund operations and ensure loans for new automobile production. Chung's entrepreneurial skills included a strategy for using and financing capital for building new businesses.

Chung Ju Young (http://www.rokps.or.kr/profile_result_ok.asp?num=1844).

What is characteristic about the founders of successful chaebols in South Korea, such as Chung, was their entrepreneurial spirit. (The second theme about change in leadership occurred in Hyundai, after Chung Ju -young's retirement, which we later discuss.)

HISTORY: HYUNDAI MOTOR COMPANY

Chung also saw a business opportunity in building automobiles. In addition to construction, ship building and equipment businesses, Chung began manufacturing automobiles. In 1967, Chung established the Hyundai Motor Company. In 1968, he sold his first car, from assembling a Ford car, called the Cortina. Again, he had needed to go abroad to find and learn the technology of automobile manufacturing.

Chairman Chung was participating in President Park's vision and plan of an export economy for South Korea. Donald Kirk wrote: "Chung's plunge into motor vehicles on a large scale ushered in a new era in Hyundai, as well as Korean history. . . The search for a foreign collaborator – a vehicle maker for whom Hyundai would assemble cars under the foreign name for local sale – found sympathy and support where it counted, in the Blue House. Park Chung Hee had in 1962 promulgated a five-year plan for motor vehicles, followed by a more specific Automobile Industry Protection Law, under which the government would promote construction of plants for building vehicles. . . . The government restricted car imports – a move that foretold the complete protection afforded by tariffs and quotas in later years." (Kirk, 1944a)

If Hyundai would build automobiles, it would have government financial assistance to build factories and a protected domestic market in which to sell autos. The first step would be to find a foreign automobile manufacturing partner for Hyundai Motor to get started in building a car – first not in making parts but in assembling a car. Chung Ju Young assigned the task of finding a partner to one of his younger brothers, Chung Se Yung. At the same time, Ford Motor Company in the U.S. was looking for a Korean partner, in order to sell Ford cars in South Korea. On February 23, 1968, Ford and Hyundai signed an agreement for Hyundai Motor Company (HMC) to assemble Ford cars in a franchise agreement. Ford would supply the knowledge of car assembly, the equipment to assemble and all the car parts for assembling 3000 vehicles per year (both cars and trucks). Hyundai would build a factory, hire and train management and workers, and provide the capital for the operation. Capital would come from profits in other companies in the Hyundai Group and in bank loans the Group could raise for the new company. This assembly experience lasted for two years through 1968 and 1969. But the Hyundai goal was to design and manufacture its own cars, not just assemble cars.

As we have seen in the other stories of successful Korean firms, technology was first acquired from outside. They bought foreign technology, learned the technology, designed its own products from the technology, and then began improving the technology. We recall this was as Youngrak Choi described: "Over the last decade, some Korean enterprises have emerged to become global players in their specialized products. How have they achieved such tremendous technological progress in so short period of time? . . . Korean firms have experienced three sequential phases . . . 'path-following', 'path-revealing' and 'path-creating'. (Choi, 2010)

Hyundai Motor was engaged in 'path-following' by assembling Ford cars. Next Hyundai wanted to launch into the 'path-revealing' phase by learning how to design and manufacture its own car. Ford wanted to own half of Hyundai Motor as it moved to build its own car. Hyundai wanted to build its operation alone. The business alliance split up. Kirk wrote: "HMC and Ford were soon bickering over management strategy, reflecting a desire of each to run the venture. . . . No matter how much they talked, or how well they got along personally, there was no getting around the basic reality. (As a Ford manager, William Hartigen expressed) 'Chung Ju-yung and Chung Se-yung had adopted the view, 'this thing is for the nation'. . . . The government in January 1973 withdrew its approval for any deal at all (between Hyundai and Ford as a joint Korean automobile venture." (Kirk,1944a)

Yet for Hyundai to design its own car, it still needed foreign technology. In 1973, Hyundai Motor signed a licensing agreement with Mitsubishi Motor. At first, Hyundai would buy Mitsubishi manufactured motors and transmissions. But Mitsubishi would give Hyundai technology for the design of the parts for a car. Hyundai would then build motors to Mitsubishi design, and Mitsubishi would buy them for some of its own cars (to lower its manufacturing investment costs, in competition with other Japanese automobile manufacturers). A few months later in 1973, President Park's government issued a policy directive for the new Korean automobile industry, directing Korea automobile makers to manufacture all the parts for their cars. Hyundai was going in the right direction to manufacture a complete car, using Mitsubishi technology.

Donald Kirk wrote: "Having already obtained the Mitsubishi technology, Chung Se-yung journeyed overseas again for still more expertise and manpower. . . . One of his emissaries was off to England, conferring with George Turnbull, a former British Leyland vice president . . . Turnbull went to work in earnest on the mission (for Hyundai) that would consume him for the next three years until he finally saw his piece of Hyundai history completed and returned home to London in 1977. 'First, we built an engineering center and got the drawings marked up in Japanese and English and got working drawings and then sent them to the toolmaker to make the tools from which they make the parts.' . . . While Turnbull was overseeing the rise o the new factory, Chung had to get the money to pay for it. . . . As the plant neared reality, Hyundai came up with a catchy name for the car it was planning, 'Pony'." (Kirk, 1944a)

Hyundai Pony 1975–1990. Assembled in Ulsan, South Korea. "George Turnbull hired five other top British car engineers, Kenneth Barnett body design, engineers John Simpson and Edward Chapman, John Crosthwaite as chassis engineer and Peter Slater as chief development engineer. With Turnbull's experience with the Morris Marina, engines and transmissions from Mitsubishi, some parts from the Ford Cortina they were already producing, and a hatchback body styled by Italian designer, Giugiaro, they developed the Hyundai Pony." (http://en.wikipedia.org/wiki/Hyundai_Pony, 2013)

The Pony was sold in South Korea and exported to Ecuador, Europe, and Canada. It was not sold in the United States, not meeting emissions. At first it was a best seller in Canada, at a reasonable price and superior in quality to then communist Eastern-bloc car imports.

In 1977, Hyundai began designing a successor to the Pony, to be called the Excel for export.. It continued to use Mitsubishi engine designs, but Hyundai manufactured the engines. The Excel was produced in 1985 and continued in production to 1994.

"The Excel was introduced as a replacement for the Hyundai Pony. In the United States it was the company's first and only model, but thanks to a price of $4,995 USD and being voted 'Best Product #10' by Fortune magazine, it set records for a first-year import by selling 168,882 units, helping push the company's cumulative production past one million by 1986." (http://en.wikipedia.org/wiki/Hyundai_Excel; 2013)

The Excel met U.S. emissions standards, and in 1985, Hyundai set out to build a sales capability in the United States. Donald Kirk wrote: "It wasn't difficult to round up veterans of the foreign car business in the United States. A number of them had been working for years for the Japanese. The Chung brothers and their American cohorts believed they could find a unique niche in car-hungry America. Hyundai with air-conditioning and automatic transmission was under $7,000. Everyone looked at this as the next Toyota." (Kirk, 1994a)

In 1985, sales first went well. Japanese car makers had agreed in 1981 to limit imports to the U.S., as then the U.S. car manufacturers were in trouble, without fuel-efficient cars. Japanese car makers would build manufacturing facilities in the U.S. to get around the import restrictions. But that would take time. Meanwhile for 1985 and 1986, Hyundai sales soared. But after 1987, U.S. car manufacturers were turning out new models; and Hyundai sales would decline, also with unfortunate quality problems. Kirk wrote: "Having dominated the compact-car league for the first year or two, the Excel now confronted competitors put out by big-time manufacturers . . Excel sales for 1988 were static at 263,000 . . plunging to 188,000 for 1989. Consumer Reports (a consumer products rating magazine) summarized the anti-Hyundai sentiment, reporting the Excel and Sonata ranked near the bottom of customer satisfaction in 1989 and 1990. By 1993, Hyundai Motor America was losing well over $100 million a year." (Kirk, 1944a)

Hyundai was in the automobile race for the long-term. Donald Southerton wrote: "Despite a decline in sales and a tarnished image, rather than dropping out of the world's largest automotive market, Hyundai would learn from the Excel. By the late 1990s and early 2000s and with leadership changes at Hyundai Motor Company, Hyundai would invest heavily in long-term R&D, design, manufacturing, and quality of their vehicles. . . . A bold '10-year or100,000 mile Warranty' was introduced in the U.S. in 1998. . . The improved quality along with gradually improving consumer perception of the cars heralded . . . a remarkable transformation in brand image and sales at record pace." (Southerton, 2012)

Hyundai had been designing its own engine to make significant improvements over the Mitsubishi design (from which it had been producing for the Excel). The new Hyundai engine was called the 'Alpha' and would prove strong and enduring (lasting over 100,000 miles in use). This was the first step toward Hyundai's 'path-revealing' phase. Youngrak Choi wrote: "Hyundai succeeded in developing the Alpha Engine in 1991, as its first independent engine. Since then, Hyundai has developed ten gasoline engines. Along the way, Hyundai developed new engines for high performance, fuel efficiency improvement, durability, quietness, and environmental standards. In 2004, Hyundai introduced a 'Theta Engine' . . . with its functionality considered to be at the same levels as the engines of Toyoda Camry and Honda Accord. The key features were aluminum alloy, metal timing belt, variable valve timing; and it was developed through a joint effort with DaimlerChrysler and Mitsubushi (with Mitsubushi adopting the engine in 2005)." (Choi, 2010)

Meanwhile at the end of the twentieth century, the Asian Financial crisis had a major impact upon restructuring the Korean automobile industry. Daewoo Motors went bankrupt and was acquired by General Motors. Samsung's car company was sold to Renault. Hyundai acquired Kia Motors. Donald Southerton wrote: "For Kia Motors, the crisis would be devastating. . . Kia found itself heavily in debt. . . By October 1997, it was clear that additional funding for the beleaguered Kia would not be forthcoming from private banks. . . The government took over the company and placed Kia in receivership in order to stave off bankruptcy and job losses. . . Hyundai eventually won the bid (for Kia) and purchased a controlling interest in its former rival, Kia Motors. Fortunately, for Kia Motors, the Hyundai Group acquisition was an opportunity for a new start." (Southerton, 2012)

The restructuring of South Korean industry in 1998 (which we will later review) began some major management changes in Korean chaebols. Southerton wrote: "For Hyundai and Kia Motors, a new management came . . . led by Chung Mong-koo, the son of the Hyundai Group founder. . .He oversaw the merging of Kia's operations with Hyundai Motor. . . Quality was set as a top priority and the management team formulated a strategy that benchmarked the world's best car brands, mandated higher production standards and declared that poor workmanship was not acceptable. . . Concurrently with the restructuring, Hyundai began production of its first SUV, the Santa Fe, . . a huge hit with the American buyer market." (Southerton, 2012)

AUTOMOBILE INDUSTRIAL LIFE CYCLE

In a commodity industry such as the automobile industry, we earlier discussed how the industrial life cycle applies. We recall that David Ford and Chris Ryan in 1981 proposed a chart of market-volume-over-time for an industry to reflect the underlying maturation of the core technologies of its product; and they called it an 'industrial life cycle' (Ford and Ryan, 1981). Next in 1985, William Abernathy and K. B. Clark described how in the early days of this pattern, the product innovations peaked and then were followed by production innovations. (Abernathy and Clark, 1985) In Figure 6.1 the industrial life cycle is compared to the rates of product and production innovations.

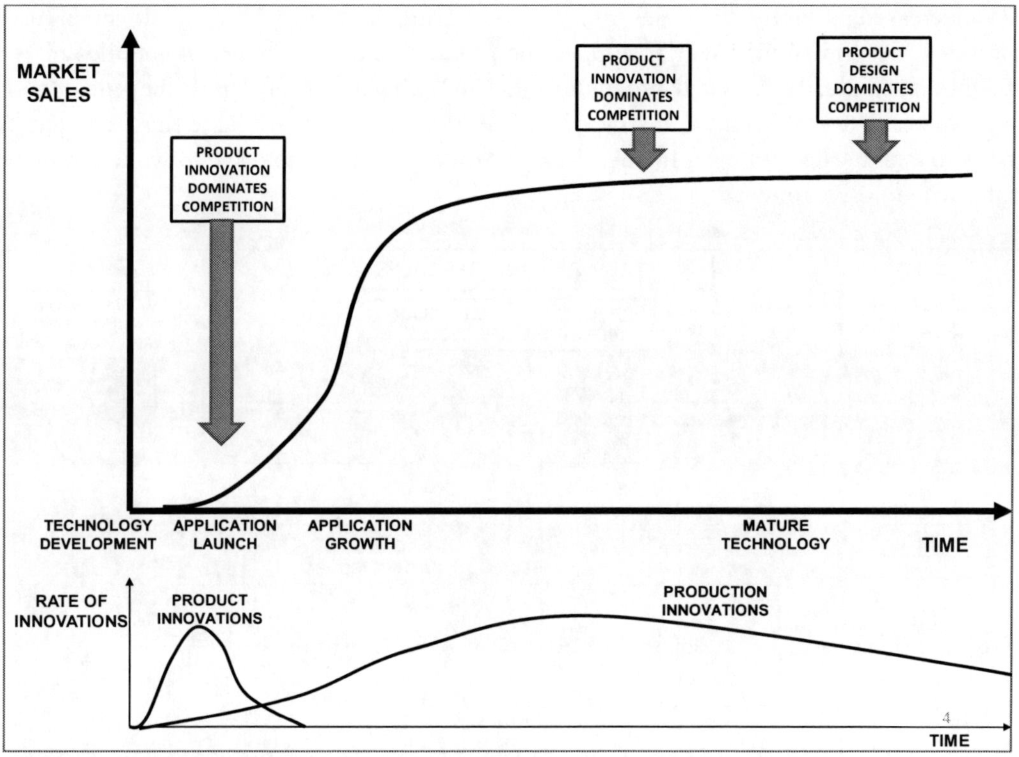

Figure 6.1. Industrial Life Cycle.

Based upon a new radical technology, new products create a new industry (e.g. the invention of the stored program electronic computer created the computer industry in 1950). In the launch phase, many new product technology innovations spur the growth of the new industry. This pattern depends upon the core technologies of the product maturing, so that product performance is no longer much improved. But after a product design-standard is achieved, the rate of product innovations decline; and innovation shifts to a high rate of production innovations, to produce the products, reducing costs and improving quality. As technical innovation both in products and production slows, then the industry enters a mature industrial phase, with commodity-like products. All products of competitors are similar in performance and cost, with differences principally in product design differentiating competition. The number of firms in the new industry increases rapidly in the launch phase to peak in the application growth phase and then decline in mature industry phase.

For competition during these phases of industry, first production innovation dominates competition during while product technology improves. Competition is over product performance. After product performance plateaus, then production technology improves and dominates competition. Competition is over product cost and quality. After the technologies in production improvement slows in progress, then product design dominates competition. Thus, in a mature product-technology industry, competition centers upon cost-quality-design.

We recall that in Figure 6.2, one can see how the data of the U.S. domestic sales of automobiles compares to the theory. The general form of the histogram follows the pattern in an industrial life cycle, with a few exceptions. First, the U.S. economic depression in 1930 to 1935 decreased abruptly the sales of cars to the dramatic rise in unemployment in the depression. Second, World War II stopped the production of consumer automobiles in 1942, until the auto industry renewed production in 1947, after the war. Third, the sales doubled from 1988 to 1998, as women in America entered the workforce in large numbers, and two cars were then needed for each family – one car for the husband to drive to work and one car for the wife to drive to work.

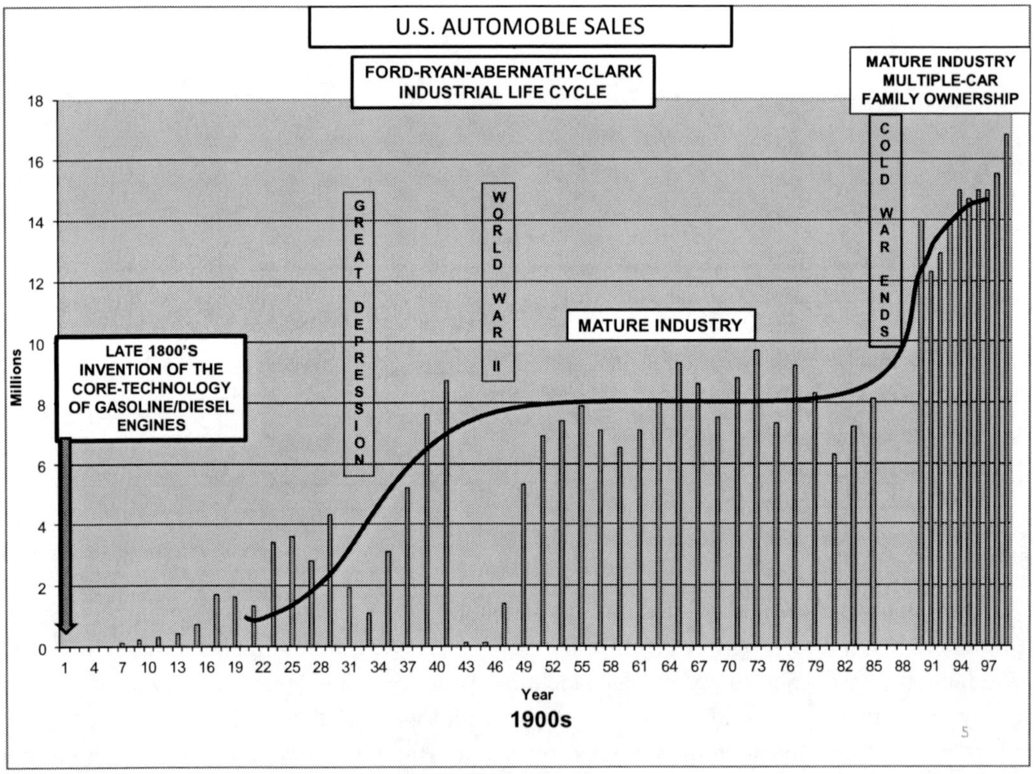

Figure 6.2. Auto Industry with One Product Core-Technology.

With these three exceptions, the industrial life cycle of the automobile industry in the U.S. is that of a core product technology (which has not much changed over the hundred year life of the industry). For example, William J. Abernathy wrote: "The year 1896 can mark the beginning of the U.S. automobile industry because then more than one auto was produced from the same plan: J. Frank Duryea made and sold 13 identical cars in Springfield, Massachusetts. During the next few years, many new automobile firms were founded and a variety of auto configurations were offered (Abernathy, 1978). Races were held between the three principle configurations of automobiles in steam, electric, or gasoline power. In 1902, a gasoline-powered car defeated electric and steam cars at a racetrack in Chicago, establishing the dominance of the gasoline engine. Thereafter, this engine was to become the core technology for the automobile. (Abernathy, 1978)

After a hundred years, the gasoline engine (or its alternative configuration the diesel engine) is still the core technology which drives cars. When Henry Ford produced his four cylinder gasoline engine for the Model T in 1918, its mileage performance was 25 miles per gallon (10.4 kilometers per liter). In 1985, when Hyundai produced the Mitsubishi-designed four-cylinder engine for its Excel, its mileage performance was still 25 miles per gallon (10.4 kilometers per liter). Of course, incremental improvements were made in the core technology of the gasoline engine, gaining more power from lighter engines. But the overall technical performance was similar. Even in 2013, a six-cylinder car attained about 25 miles per gallon of the road and a four-cylinder car attained 35-39 miles per gallon. The efficiency of the gasoline engine was bounded by the physics of the thermal cycle of the engine, and that physics had not changed.

The physics underlying the basic chemistry of the battery changed had not changed either. Back in 1902, a battery-powered, electrical car could only go a few miles without recharge; and in 2013, a lead-acid battery electrical car could still only go about 40 miles without recharge (on a level road without hills). Also in 2013, a lithium-ion battery powered electrical car (with the batteries taking up the whole of the base of the chassis of the car) could still only go 300 miles on a single charge, on a level road.

Thus the gasoline/diesel engine remained and would likely remain the core technology of the automobile industry. What did change in the automobile industry was the technology of design. It was this technology which provided Hyundai Motor with the opportunity of becoming a global competitor.

As Youngrak Choi had described, Korean development in Hyundai did occur in the stages of 'path-following', path-revealing' and 'path-creating'. And in each stage, Hyundai's technology progressed: "The stages of technology capability building in the Hyundai Motor Company can be divided into the stage of assembly manufacturing (1967-74), the stage of establishing independent brand (1975-1990), and the stage of self-generating technology (1990-present). Technology importation was the key source of technology development in the early stages. Path revealing began in 1991 when Hyundai succeeded in developing the Alpha engine, as its first independent engine." (Choi, 2010)

In the early stages of technology importation, Hyundai had to import an automobile for assembly from Ford to learn the technologies of automobile assembly and to build the manufacturing facilities to build a car. In the next stage of building its first car, Hyundai had to import the engine and drive train from Mitsubishi, while building the car body and other components to build its first car, the Pony. In the subsequent stage, Hyundai manufactured its own engine and drive train for its next car, the Excel, importing Mitsubishi engine designs. Afterwards in generating its own technology, Hyundai developed and designed its own engines, starting with the Alpha engine.

For a complicated physical product, such as an automobile, technology learning and acquisition took time, from 1967 to 1991. At the same time Hyundai was earning revenue from producing and selling cars. What is striking in Chung's entrepreneurial story of Hyundai building a car company was:

(1) The willingness of Hyundai to seek out and find technology alliances with foreign car manufacturers;
(2) The long-term plan for Hyundai to build technology and manufacturing capabilities to become a globally competitive automobile company;
(3) The capital capacity to make continual investments in technology and manufacturing and operations both to learn and succeed commercially.

These were the three principal factors in Hyundai Motor's success – long-term commitment, long-term planning, long-term capital capability.

From the beginning of Chung's founding the Hyundai Group, his experience was of operating in a global context. Chung grew up in a foreign power occupied Korea. Korea was divided in a global context of the two super-powers of the U.S. and the U.S.S.R. The Korean war was a civil war begun and fought with global participants, i.e. China, the U.S. and United Nations. Chung's post war businesses were built upon global markets, first the U.S. Army and next Middle East countries. When Chung began Hyundai Motor in 1967, Chung and Park had a long-term strategy of building an export car industry for a global market. And in carrying out the car strategy, Hyundai systematically, by stages, acquired and learned technology from global company sources – Ford, British Leland, Mitsubishi.

The Third Stage: Path: Creating in Hyundai Motor Company

From the technology perspective, product performance technology was relatively mature and production technology for manufacturing an automobile was also relatively mature. But there was a technology revolution in design going on globally, just as Hyundai was entering its 'stage of self-generating technology'. Youngrak Choi wrote: "They have also been active technological learning activities such as . . . introduction of advanced design and production technologies like CAD,CAM, 3D design and simulation, and the establishments of R&D institutes." (Choi, 2010) We recall that in a commodity industry (wherein the core technology is mature), then the competitive edge is not in product innovation but in product cost, quality, and design. For Hyundai, to begin path-creating the technology opportunity lay in a revolution in design technology. Briefly we review the technological progress in computerization of design and production from the 1980s: CAD, CAE, CAM, CIM.

CAD/CAE – Computerized Design Technology

After the electronic stored-program computer was invented from 1946, the application of computers to information (information systems) and communication (Internet) has been extensive. Computers had also been applied to design and manufacturing. In design, computer-assisted design technology has been called Computer Aided Design (CAD). In manufacturing, computer-assisted manufacturing technology has been called Computer Aided Manufacturing (CAM). The growth of Hyundai Motor in the two decades of 1990s and 2000s was just at the right time to take advantage of these for the design and manufacture of Hyundai automobiles.

CAD software for mechanical design shows the geometry of mechanical objects, all pictured by the computer. It is used in the design of objects and production of engineering drawings by which to make mechanical or electronic things. It pictures things in two-dimensions (2D) or in three dimensions (3D), often called 'solids modeling'. CAD is a computational design technology used in the design of automobiles, airplanes, ships, appliances – any physical thing which can be manufacture. Figure 6.3 shows how engineering drafting drawings appear in a computerized CAD screen.

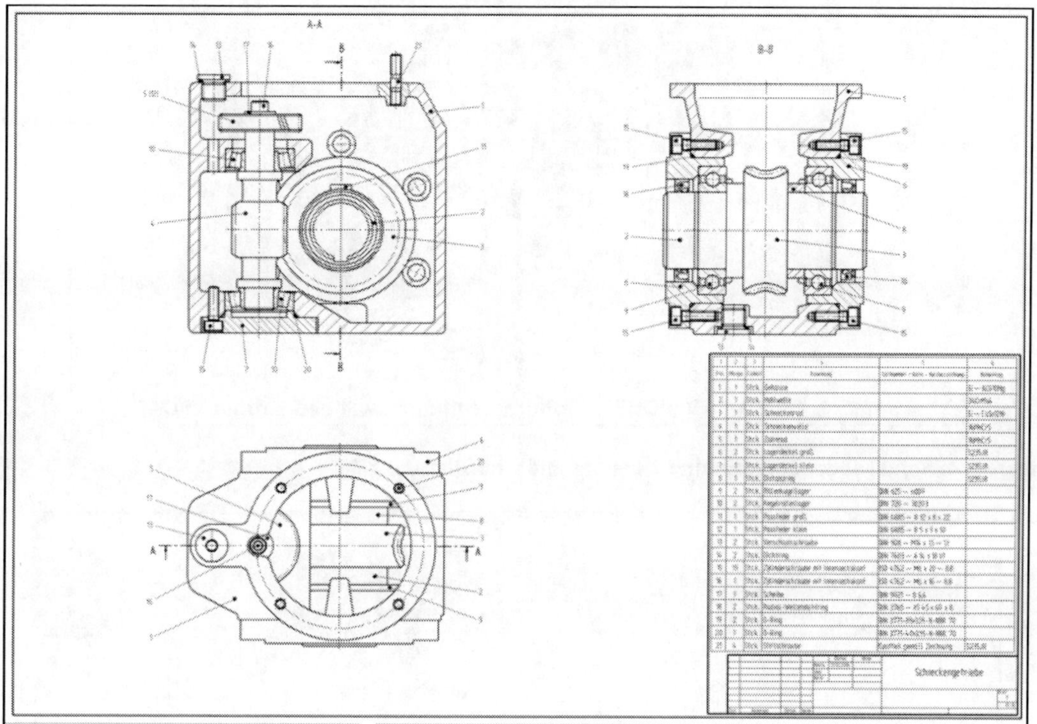

2D CAD drawing   Source: (http://en.wikipedia.org/wiki/CAD)

Figure 6.3. CAD – Drafting Drawings of a Mechanical Part.

CAD software history traces back to Ivan Sutherland, when in 1963, he had been doing his PhD research at the Massachusetts Institute of technology. He wrote a software program in which a designer could interact with a graphic representation of the geometry of a product design. Afterwards, Sutherland founded a company and commercialized this product, he called Sketchpad. In 1971, P. J. Hanratty wrote a software program, which he called Automated Drafting and Machining, which he sold to many companies. Other companies developed and improved software in which to graphically design the geometry of a hardware product. These included McDonnell Douglas (Unigraphics), Computervision (CADDS), Calma, Gerber, Autotrol and Control Data. By 1981, computer-aided design software had progressed to displaying solid-looking 3-D models, such as Romulus (ShapeData), PADL-2, Uni-Solid (Unigraphics), CATIA (Dassault Systemes), Autodesk (founded by John Walker).

The first commercial airplane to be designed entirely by CAD was the Boeing 777, using the CATIA software. Figure 6.4 shows a solid model rendition of the pistons in an engine.

3D CAD MODEL Source: (http://en.wikipedia.org/wiki/CAD)

Figure 6.4. Solid Model of an Engine's Pistons and Cranshaft.

FEA mesh created by an analyst prior to finding a solution to a magnetic problem using FEM software.

Source: (http://en.wikipedia.org/wiki/Finite_element_analysis)

FEA solution to the problem, involving a cylindrically shaped magnetic shield. The ferromagnetic cylindrical part is shielding the area inside the cylinder by diverting the magnetic field created by the coil (rectangular area on the right). The color represents the amplitude of the magnetic flux density,

Figure 6.5. Finite Element Analysis.

However, in engineering design of a physical thing, one not only must specify the geometry but also the physics of the thing – not only how it looks but how it operates. The important addition to CAD was another computational technique called Finite Element Analysis (FEA). CAD does geometry, and FEA does physics. In the geometry of a thing designed in the computer, FEA software constructs a grid of points all over the shape and inputs the physical forces acting at each point in the object.

As forces are applied to the overall thing, then the computer can calculate what kinds of strains and distortions can occur in the thing. This software embodies physical equations and theories which predict how materials flow or deform under forces. These equations the basis of the 'algorithms' in the software. (They include physical theories such as the Euler-Bernoulli beam equation, the heat equation, or the Navier-Stokes equations – all being used to express changes in small areas of the physical system of the designed object.). Figure 6.5 shows a computerized FEA picture of the physics of an thing designed to be a magnetic shield, with the lines showing the field of magnetic forces. The first picture is a computer-generated mesh upon which to compute the magnetic field around the object. The second picture shows the results of the computation, the magnetic field around the object.

When one designs a physical product in CAD and then performs CAE analysis on it, one can predict how the physical product will react under forces in the operation of the product. For example, CAD/CAE has been used to improve the safety of automobiles by predicting how the automobile will deform in a car crash, Figure 6.6.

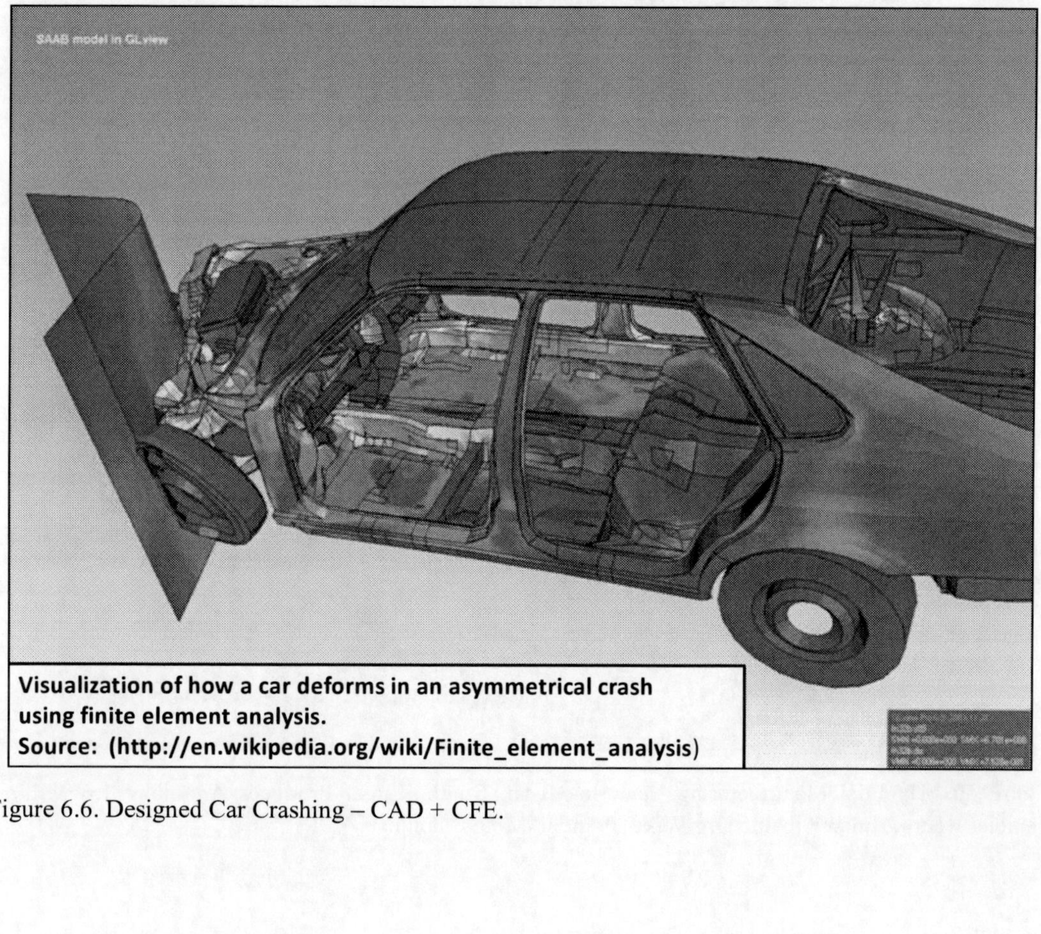

Visualization of how a car deforms in an asymmetrical crash using finite element analysis.
Source: (http://en.wikipedia.org/wiki/Finite_element_analysis)

Figure 6.6. Designed Car Crashing – CAD + CFE.

Computerized Manufacturing Technology – CAM/CIE

In the 1980s, computerized technology not only impacted design but also manufacturing, and the computer could connect design to manufacturing. This was called CAD/CAM, as computer aided design and manufacturing. Earlier in the 1970s, controls on machine tools, such as lathes or milling machines, were being converted from human hand control to computer control, and this new conversion was called 'numerical control'. This was an automation of machine tools by operating them from instructions from a computer. In 1952 at the Massachusetts Institute of Technology, a milling machine was altered with electrical motors that moved the control of the machining operation. At first, these motors were directed by numbers punched into tapes, and then later directed by computer instructions.

Computer Aided Manufacturing (CAM) uses numerically-controlled machines, so that instructions from a computerized design can be sent to a computer-controlled machine tool to machine the features of the designed product. This kind of computer-controlled automation extends to all manufacturing processes, from machine tools to mold-injection machines. Computer control also controls the production processes of whole production systems, such chemical production or steel making (as in Posco).

The linking of information from computerized design to computerized manufacturing to running an entire computer aided enterprise has also been given designated acronyms, such as computer-integrated manufacturing (CIM) and computer-integrated enterprises (CIE).

As Hyundai Motor was developing its own technology in the 1990s, new generations of Korean engineers were being trained in Korean universities in the new computerized design and manufacturing technologies. These new skills were used by Hyundai to move toward its 'Path-Creating' stage. Figure 6.6 shows part of the assembly lines at Hyundai in Ulsan.

Figure 6.6. Hyundai Manufacturing Plant In Ulsan, South Korea, Produces Annujally 1.6 Million Vehicles. (Http://En.Wikipedia.Org/Wiki/Hyundai_Motor_Company).

HISTORY: SUCCESSION AT HYUNDAI

We have emphasized how well Chairman Chung of the Hyundai Group and President Park of the Republic of Korea got along and shared a vision for a strong and prosperous South Korea. And we noted how after Park's assassination, a younger general Chun Doo-hwan took power and became president under Park's legacy of the Yushin constitution. Chung in industry had next to formulate a strategic alliance with Chun in government. Donald Kirk wrote: "The Chun regime . . . was working on measures ot improve competitiveness and efficiency, all in accordance of the views of military and academic theorists rather than practitioners. . . One of Chun's aides believed they could trim Chung – and the chaebol – down to size. . . For Chung, the trick was how to win Chun's favor . . . . Chung had time on his side – that and the acquired taste among Chun and his allies for a share of the riches of their new business acquaintances." (Kirk, 2004a)

Later in 1997, long after President Chun had left office, a new government, under Kim Young-sam, accused the former President Chun of corruption. For example in 2013, Choe Sang Hun wrote: "State prosecutors equipped with metal detectors raided the Seoul residence of former President Chun Doo-hwan on Tuesday in a search for assets. Mr. Chun, the former military dictator, owes South Korea 167.5 billion won, or $150 million, in fines but claims to be broke. In a Supreme Court ruling in 1997, Mr. Chun, now 82, was ordered to return to the state 220 billion won he had illegally accumulated through bribery from big businesses during his eight and a half years in power in the 1980s. He has so far paid only a quarter of the amount. . . . Mr. Chun has rarely appeared in public since he stepped down in 1988 and entered a Buddhist monastery. In the 1997 verdict, he was also convicted of sedition for his role in the 1979 military coup that brought him to power and a 1980 military crackdown that left hundreds of people dead in the southwestern city of Kwangju. He initially was sentenced to death, but the penalty was reduced to a life imprisonment. He was later pardoned and freed. On Tuesday (2013), a team of 90 prosecutors, tax collectors and other investigators ransacked Mr. Chun's home in eastern Seoul. Television footage showed them hauling away paintings, porcelain and expensive artifacts." (Choe , 2013a)

The relationship of the chaebols to the two military presidencies (after Park) appearently involved donations to political campaigns and to charities associated with politicians. Donald Kirk wrote: "In the swing to democracy in 1987 (after Chun and under Roh), pressure mounted for a full-scale expose of corruption in the Chun administration. . . Chung was compelled to appear as a reluctant witness . . . during nationally televised parliamentary hearings on his role as chairman of the Ilhae Foundation, established by Chun in 1983. (Chung testified) 'I knew that the foundation was originally set up to provide financial assistance to the bereaved families of the people who died in the Rangoon bombing attack (of South Korean officials by a North Korea spy). So I thought that 10 billion won was enough for the establishment of Ilhae. . . We could not help donating additional money to prevent unfavorable treatment.' Ilhae amassed 60 billion won in donations." (Kirk, 1944a)

But after Roh's administration, Chung began to think he needed to enter politics. Kirk wrote: "Just as Chun Doo-hwan at the outset of his rule had been attuned to the idealistic advice of theorists and planners, so Roh on succeeding Chun in 1988 had shown signs of wanting to curb the chaebol. . . . Chung Ju-yung soon lost patience. He believe government was 'stupid' to force Hyundai to sell off land set aside for office buildings, said aide Lee

Byung-kyu, citing the real estate regulation as 'an example of wrong government policy.' . . . Chung found much else to criticize, castigating Roh in speech after speech for 'misspending' and 'waste' and 'abuse'. He had had enough of generals and their policies and plans. If they believed they could run his business better than he, who was to say he could not run the country better than they?" (Kirk, 1944a)

For Chung, the difference between the General Park's government and the Generals Chun's and Roh's governments was this. General Park never interfered with the conduct of businesses nor accepted 'donations' from business leaders. Chung got along well with a general who supported chaebols as long as they were competitively successful and exported for the benefit of Korea. And Chung got along well with a general who was incorruptible. President Park, had his dictatorial faults, as in the Yushin constitution and in the KCIA brutal suppression of civilian rivals, such as Kim Jae-jung. But Park had supported the entrepreneurship of capable businessmen, and he hadn't accepted bribes.

In the 1992, Chung formally resigned as chairman of the Hyundai Group and ran for president against Kim Young-sam. Donald Kirk wrote: "Chung cast his campaign as a crusade against corruption. . . In self-styled 'retirement', Chung had discovered an exciting way to keep busy. . . In July he moved his party headquarters . . . to an office tower owned by Hyundai Engineering and Construction . . in the heart of the capital. . . No matter how hard he claimed to have retired entirely from Hyundai, Chung fell back of Hyundai money and contracts. . . . In the campaign's last week, all but Chung and his yes-men could see the obvious: he had misled himself. . . (At the end of the election day loss) Chung said: "I accept the results of the election. . . I am sending a congratulatory message to Kim Young Sam." (Kirk, 1944a)

Chung then did retire, while still retaining control over Hyundai, in a Confucian manner. Kirk wrote: "By the beginning of 1994, four or five days a week, the old man tended his farm on the Korean west coast on Sosan Bay. It was no ordinary farm, however, and he was no ordinary farmer. The farmer was Chung Ju Yung, Korea's richest man, founder of the mighty Hyundai Group. As a major actor in this Northeast Asian drama, Chung Ju-yung had as much as anyone turned South Korea into an industrial nation." (Kirk, 1994a)

After Chung's death in 2001, the Hyundai Group was re-organized and power passed to brothers and sons.

Brothers

> "After leaving the Hyundai Group, the brother, Chung In-yung, founded the Halla Business Group, whose interests, included Mando Machinery, Halla Cement, Halla Construction, Halla Heavy Industries, Halla Climate Control Corp. After working for Hyundai Engineering & Construction, Chung Soon-yung. Left and (taking with him Hyundai Cement) formed the Sungwoo Business Group (which includes Hyundai Cement, Hyundai Welding, Sungwoo Automotive). The brother, Chung Shin-yung died in an auto accident in Germany while working as a journalist for a Korean newspaper company. His only son Chung Mong-hyuk ran Hyundai Oilbank, the third largest oil refiner in Korea. The brother Chung Sang-yung was founder of the KCC Chemical (Keumkang) group, Korea's leading paint and glass maker. The brother Chung Se-yung helped found Hyundai Motor and later left the Hyundai Group with the Hyundai Development Co., Ltd. (a leading housing builder in Korea)." (http://en.wikipedia.org/wiki/Chung_Ju-yung; 2013)

Sons

"Chung Mong-pil. Died in a car accident in Ulsan. Chung Mong-koo became head of the Hyundai Kia Automotive Group. Chung Mong-kun became Chairman of the Hyundai Department Store Group, one of the largest retailers in South Korea. Chung Mong-woo. committed suicide leaving three sons. His oldest son Chung Il-sun became the President of BNG Steel, a member of the Hyundai Kia Automotive Group. Chung Mong-hun was former Chairman of the Hyundai Group and heir apparent to his father. He committed suicide in August 2003. Chung Mong-joon was a politician and de facto owner of Hyundai Heavy Industries, the world's largest shipbuilding firm, as well as Vice Chairman of FIFA. Chung Mong-yoon was chairman of Hyundai Marine & Fire Insurance, Korea's third largest non-life insurer. Chung Mong-il was Chairman of Hyundai Merchant Bank and Kangwon Bank." (http://en.wikipedia.org/wiki/Chung_Ju-yung; 2013)

In 2013, this was a summary of the status. "The Hyundai Motor Company is a South Korean multinational automaker headquartered in Seoul, South Korea. The company was founded in 1967 and, along with its 32.8% owned subsidiary, Kia Motors, together comprise the Hyundai Motor Group, which is the world's fifth largest automaker based on annual vehicle sales in 2012. As of 2010, the Company sold over 3.6 million vehicles worldwide. Hyundai operates the world's largest integrated automobile manufacturing facility in Ulsan, South Korea, which has an annual production capacity of 1.6 million units. The company employs about 75,000 people worldwide. Hyundai vehicles are sold in 193 countries through some 6,000 dealerships and showrooms. (http://en.wikipedia.org/wiki/Hyundai_Motor_Company; 2013)

LESSONS LEARNED

We have seen in the development of Hyundai that – entrepreneurship, capital, strategy, and technology, and globalization – all were important factors. Chung Ju-young was an entrepreneur always looking for opportunities to start new businesses. When seeing a new business opportunity, Chung used existing capital accumulation within his chaebol and sought out loans to finance his new venture.

Chung always operated within a global context, finding business opportunities internationally and focused upon exporting Hyundai products and operations. Chung also operated within a global financial context, but had sufficient cash within his chaebol, so as not to face bankruptcy when international financial conditions suddenly went unstable.

To profit from a new business, Chung's strategy was always to deliver a product quickly and at a price below competitors. To produce a product, Chung always searched for the best technology from foreign competitors, licensing the technology and using foreign managers to learn how to build a new business' operations. Chung focused his technology acquisition upon 'path-finding' and 'path-revealing' in learning and developing it for his own product designs and machinery.

Successful entrepreneurship needs to be focused on long-term strategy for finding business opportunities, accessing capital, acquiring competitive technology, acting globally. Leaders in industry – who are bold, competent and honest – build economic growth.

Chapter 7

GLOBAL FINANCES AND NATIONAL DEVELOPMENT – ASIAN CRISIS

INTRODUCTION

One of the important conditions for economic development in industrializing nations is global finance. Initially, the indigenous capital within the industrializing nation is insufficient for the capital needs of economic development. The history of all national industrializations has required capital from external finance or trade to fund the costs of building manufacturing industries and economic infrastructure. For example, in the United States industrialization in the 1800s, British capital played a major role, particularly in constructing the U.S. railroad industry. We next examine how the global financial system played an important role in the growth (and fragility) of the development of the Korean economy, Figure 7.1.

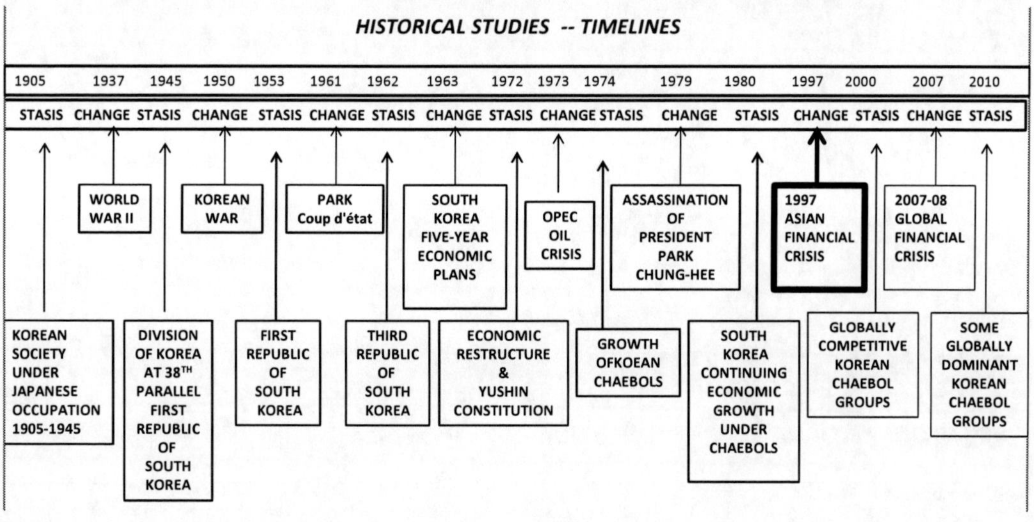

Figure 7.1. Change Events in Republic of South Korea.

1997 Asian Financial Crisis

A crisis in global finance in 1997 affected not just Korea but many other countries in Asia and else, Figure 7.2 It was a crisis point because of the massive withdrawal of capital from the 'emerging' economies of Asia. Back in 1987, financial funds in the U.S. had begun investing in the 'emerging markets'. Paul Blustein wrote: "Mark Mobious, head of the merging-markets division for Templeton funds . . loved to tell . . . (about) when his first emerging markets fund had started in 1987, it manged just $87 million. . . . Ten years later, the amount of money . . had risen more than one-hundredfold. . . . By spring 1995, enthusiasm for emerging markets had spread so widely that nearly 1000 mutual funds were active in them." (Blustein, 2003)

Not only were mutual funds providing large amounts of capital to the emerging countries, but so also were banks. Blustein wrote: "The chief draw was Asia, where the biggest providers of foreign capital weren't portfolio managers like Mobius, but banks . . . The dominant players in Asia were Japanese banks. . . . In all the emerging market combined, not just in Asia, the net inflows of private capital . . . soared to $329 billion by 1996." (Blustein, 2003) Japanese banks and U.S. mutual funds were investing capital in Asian countries.

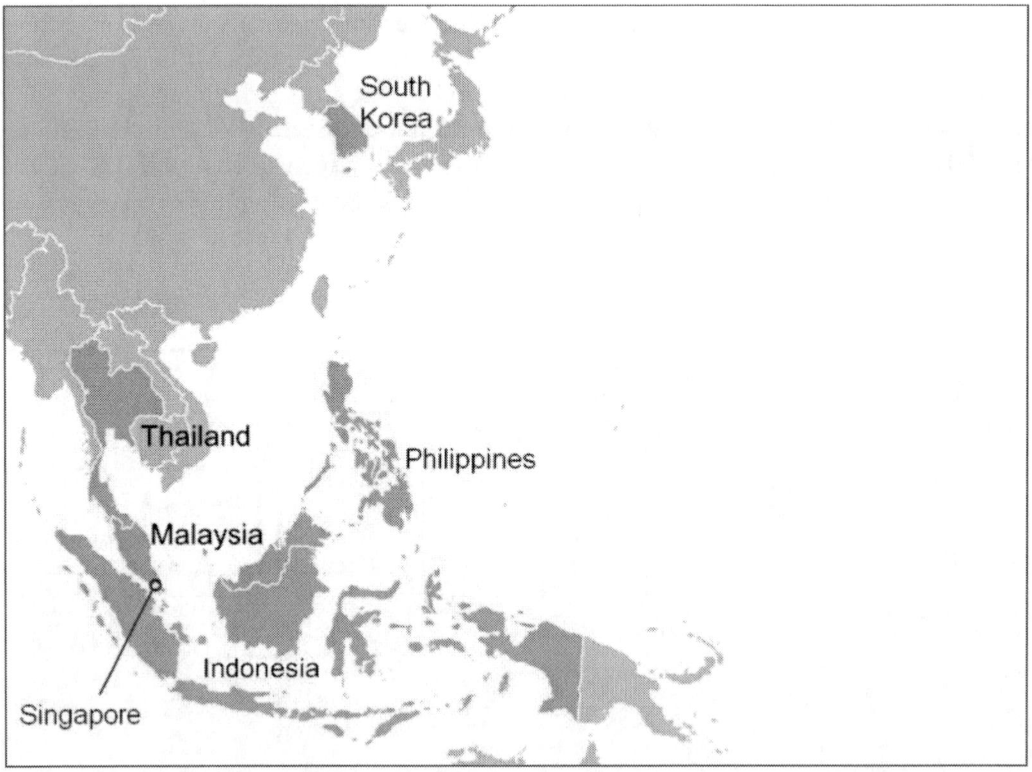

Figure 7.2. The Countries Most Affected By the 1997 Asian Financial Crisis. http://en.wikipedia.org/wiki/Asian_financial_crisis.

The crisis also affected the countries of Philippine and the territory of Hong Kong. However, we will focus on the crisis in Thailand, Indonesia, and South Korea. The first two preceded and directly influenced the crisis in South Korea.

FINANCIAL CRISIS IN THAILAND – JULY 1997

The first Asian 'emerging market' to financially stumble in July 1975 was Thailand. Blustein wrote: ". . . amid severe financial turmoil, the Thai government abandoned its long-standing policy of maintaining a fixed value for the baht against the U.S. dollar, and the baht was sinking fast. . ." (Blustein, 2003) When such an exchange rate drops, such as the baht against the dollar, then capital in baht money declines in value against capital in dollars. Then capitalists holding bahts lost wealth, and capitalists holding dollars didn't lose wealth. The Thai government tried to defend the baht by using dollars in its Central Bank's reserve funds to buy baht, until it ran out of dollars.

In Thailand in the late 1980s, investments had been made in manufacturing clothes, shoes, and some auto parts. Inexpensive labor made their prices of goods cheap to export. This stimulated a construction boom in buildings and apartments for a newly emerging Thai wealthy and middle class. Finance companies were begun, in addition to banks. For example, Finance One was established by Pin Chakkaphak, who had graduated from a business school in the United States. By 1996, Thai financial companies were providing 20% of the credit in Thailand, with Thai banks providing 80%. Financial companies were not only funding real estate investments but also stocks purchased on the margin in the growing Thai stock market. To fund this real estate boom and stock market boom, Thai banks and finance companies borrowed short term from Japanese banks and U.S. investment funds.

Financial bubbles usually burst, as did the Thai real estate and stock markets: Blustein wrote: "By early 1997, Thai banks and finance companies were reporting increases in their non-performing loans . . to 12 percent of all loans. . . ." (Blustein, 2003) For example earlier in 1993, the Bangkok Bank of Commerce had revealed that over 40% of their loans were bad. In March 1997, Finance One went bankrupt and was merged into a commercial bank, Thai Danu. (Later in July, insolvent banks would be closed.)

In the international finance socio-technical system, when financial institutions within a country begin to fail, then international currency traders start to discount the country's currency. Paul Blustein wrote: "The fame of hedge-fund managers derives from . . . predation, shooting down currencies and markets. . . . Given their reputation for canniness, hedge funds can help ignite or accelerate a run on a currency by betting on its fall." (Blustein, 2003) And hedge funds began shorting the Thai baht. One of these was George Soros' famous fund. George Soros wrote: "By selling the Thai baht short in January 1997, the Quantum Funds managed by my investment company sent a signal that it may be overvalued." (Soros, 2004).

For buying baht, the Bank of Thailand, Thailand's central bank, had $28 billion dollars In January of 1997, with which to defend the ratio of the baht to dollar. But by May, the bank had spent $23 billion to defend the baht, having only $5 billion dollars left in the bank's account. This low reserve, the Bank kept secret, unknown to the public. But finally, the Bank ran out of dollars. Blustein wrote: "On May 15 . . . The Bank of Thailand ordered all twenty-nine local and foreign banks in Thailand to stop lending baht to anyone outside the country. . . .This move inflicted acute pain on the hedge funds, to the tune of $400 million to $500 million in losses. Now it was virtually impossible for anyone to short the baht since, going short, required borrowing Thai currency. . . ." (Blustein, 2003)

The U.S. hedge funds had earlier borrowed Thai currency to sell the currency short. Later to fulfill their currency shorts, then they had to buy Thai currency from nations other then

Thailand – at an enormously inflated cost. Blustein wrote: "Sanley Druckenmiller, Soros's chief lieutenant would later admit: 'They kicked our butt.'" (Blustein, 2003)

But this did not end the crisis, since Thai banks and finance companies had borrowed a large amount of money in short term loans from international financial markets. International banks and funds now called these short term loans and refused to renew them, refused to 'roll-over' the loans, because of the shakiness of the exchange value of the baht.

Next on July, 2, 1997, the Thai government decided it had to float the baht's international exchange rate. Then the Thai government began negotiations with the International Monetary Fund (IMF) for a loan to the country to bail it out of its capital crisis. In August, the IMF board approved a loan to the government (totaling $17 billion). In return, the government promised reforms to its financial system, including closing down insolvent financial institutions. On August 5, the Thai government closed 58 financial institutions. The closure of the insolvent banks left many Thai businesses without financing. These would then also close, beginning the recession in Thailand.

And in September, because of the public requirements of the IMF loan, the Bank of Thailand had to announce how few dollars it had left. A new run on the baht began, with the baht tumbling from 25 to the dollar toward 56 to the dollar, half its previous international value.

Thus in 1997, Thailand's construction boom had ended, and many jobs were lost in construction, finance, and real estate. The stock market dropped 25% off its high back in 1997. Only after 1999 through 2001 did Thailand's economy began slow recovery.

Financial Crisis in Indonesia – July 1997

The term 'financial contagion' is used to indicate a financial crisis which spreads from one institution or country to another institution or country. The Asian financial crisis in 1997 was a contagion, spreading from Thailand in July to Indonesia in October 1997. Paul Blustein wrote: "On October 8, 1997, . . an IMF mission was dispatched to negotiate a program, including a substantial loan. . . . The Indonesian rupiah . . was starting to drop sharply." (Blustein, 2003)

Under the Suharto government, some economic growth had occurred. The lives of Indonesians had significantly improved, but is capitalist economy was still skewed by corruption. Blustein summarized this: "The country's most glaring problem was 'KKN' – Korupsi, Kolusi, dan Nepotisme (corruption, collusion and nepotism) . . . . By the 1990s, the Suharto family's avarice was so pervasive that almost any foreign firm investing in, say, a power plant or phone system or petrochemical factory had to hand over lucrative partnership rights to one presidential relative or another to grease the project's way through the country's bureaucracy." (Blustein, 2003)

The corruption increased an imbalance of wealth. The second son of Suharto headed a group, which held a television license, the right to overseas distribution of Indonesian oil production, and a tariff-protected plastics plant. Another son had a monopoly upon the distribution of cloves (used in scented cigarettes in Indonesia) and government funding to build a 'national car'. The daughters benefited, as their companies received government contracts to build power plants and toll roads.

This corruption was so large and pervasive that it weakened the financial system of Indonesia. In 1996, a World Bank team had gone to the country to evaluate its banks. The Indonesian banks had funneled loans to Suharto favored businesses, so that bad loans in the banks exceeded 25% of total loans. Blustein wrote: "The central bank, Bank Indonesia, was turning a blind eye to these sorts of activities – presumably in exchange for bribes. Foreign and domestic bankers confided horror stories and market scuttlebutt about Bank Indonesia examiners taking under-the-table handouts." (Blustein, 2003)

In July and August 1997, the rupiah's exchange rate continued to drop; and on August 14, the central bank announced that the rupiah would float on international exchanges – no longer fixed to the dollar. There was an economic problem about this float. Paul Blustein wrote: "Like their Thai counterparts, Indonesian corporations had borrowed heavily from foreign lenders . . (And) Indonesia's capital markets had drawn a rising tide of money from aboard, averaging 4 percent of GDP annually." (Blustein, 2003) When the rupiah fell, all these foreign loans were recalled, and by September 1997, the financial crisis accelerated. The rupiah had been worth 2400 to the dollar in August but fell to 3400 to the dollar by October 1, 1997 – a fall in monetary value of 42%.

So in October 1997, a team from the International Monetary fund (IMF) arrived in Indonesia to prepare a rescue loan for the Indonesian central bank. As conditions for the loan, the IMF team pushed for financial reforms. The IMF program would provide a total of $33 billion to the central bank. In return, there would be some reforms. The Bank of Indonesia closed sixteen insolvent private Indonesian banks (including one with one quarter ownership by the son of the President of Indonesia).

But in November, because of worries about the health of the other private banks, a run on many private banks began. Savers moved their money to state-owned banks. The financial crisis continued to deepen, and the rupiah rose to 5000 to the dollar on December 12th. 1997. And the rupiah continued to rise in January 1998 to 11,000 rupiah's per dollar (a fall in value of 358%).

In February 1988, the Governor of Bank Indonesia was fired. In May 1998, President Suharto resigned. President B. J.Habibie replaced Suharto as president. In October 1999, Abdurrahman Wahid became the new President. In January 2000, the government and IMF implemented a new loan and program of recovery. With reforms, the government acquired non-performing loans from the banks and restructured the debt of corporations. By 2004, six years later, the economy finally recovered.

FINANCIAL CRISIS IN SOUTH KOREA – NOVEMBER 1997

The contagion of the Asian financial crisis was not over. Paul Blustein wrote: "A likely explanation for the contagion is financial – that is, the tendency of lenders and investors who find themselves facing losses in one country to start reducing their holdings in similar countries. . . (And there is the) herding behavior in which portfolio managers follow their competitors' investing patterns lest they be blamed for failing to spot a trend that was obvious to everyone else." (Blustein, 203)

But not only were trading funds involved in herding behavior, so were bankers. Banks make long-term loans, but also short-term loans mostly to other banks. Normally, short-term loans between banks are rolled over. But in a financial crisis, short-term banks loans are not renewed and are instead called in. Carmen Rienhardt and Graciela Kaminsky wrote: "People tend to think of 'hot money' as mutual funds, or hedge funds, which of course it is. But in Asia, bank lending was the key form of hot money. . . . As regards the propagation of shocks across national borders during the Asian crisis, the behavior of foreign banks, particularly Japanese banks that began drastically to curtail their lending to the affected Asian countries following the Thai devaluation, appears to have played a role in spreading the crisis, particularly to Indonesia, Malaysia, and South Korea. The large exposure of European banks to South Korea and their subsequent retrenchment further deepened the regional liquidity crunch." (Reinhardt and Kaminsky, 2001)

In South Korea, under Park's export growth strategy, Korean chaebols had borrowed large sums of foreign money to finance expansion. Korean chaebols were heavily indebted. Myung Hun Kang wrote: "Korean firms in general are highly leveraged. The debt-equity ratio (in 1996) of the average manufacturing firm was about 4 to 1." (Kang, 1996, p 213) This is very high for corporations. A ratio of 1 to 1 (equal amounts of equity and debt) is considered conservative finance for production organizations.

In 1997, the high debt began to overwhelm some chaebols, and several went bankrupt. The government of President Kim Yung-sam refused to bail out poorly performing chaebol. In January 1997, the Hanbo steel group went bankrupt. The Sammi and Jinro and Kia groups went bankrupt But these were few compared to the other many chaebols still prospering. Thus it was not the Korean economy (unlike the Thai economy) which was collapsing. It was the contagion by the international financial community which was to collapse the Korean money, the exchange rate of the won to the dollar.

Protection of the won was the strategy of the central bank of Korea, buying won on the open market with dollars in their bank reserves. Paul Blustein wrote: "Only a couple of weeks before in (October), in conversations with IMF officials, the Koreans had put their reserves at $24 billion . . . Now (in November) the Bank of Korea staffers were citing figures (of) $9 billion and declining at a rate of roughly $1 billion a day. This was mainly because foreign banks, which had previously made short-term dollar loans to Korean banks, and routinely extended them month after month, were suddenly demanding repayment as the loans came due. . . . Short term debts owed by Korean firms to foreigners totaled about. . . $120 billion. . . . The IMF would immediately disburse $5.6 billion, and another disbursement would follow two weeks later . . . . But the Electronic Herd showed little sign of being impressed by the Funds rescue efforts, and within days Korea was in even worse financial straits than before." (Blustein, 2003)

The won continued to fall, down 49% by the end of December. The $9 billion the IMF had loaned to the South Korean central bank had gone out to foreign banks. Also in December 1997, elections were held and Kim Dae-jung became the President of the Republic of South Korea. His policy was for economic reform and restructuring. Hyundai Motors acquired the bankrupt Kia Motors. Samsung Motors was acquired by Renault. General Motors acquired Daewoo Motors. The year 1998 was a hard year in South Korea, with its economy shrinking 5.8 percent. But reforms and restructuring of Chaebol finances resulted again in the economic growth of 19%, a year later in 1999.

SOCIETAL PERCEPTUAL ANALYSIS OF THE ASIAN FINANCIAL CRISIS

We can analyze this crisis in a general societal perceptual space, Figure 7.3.

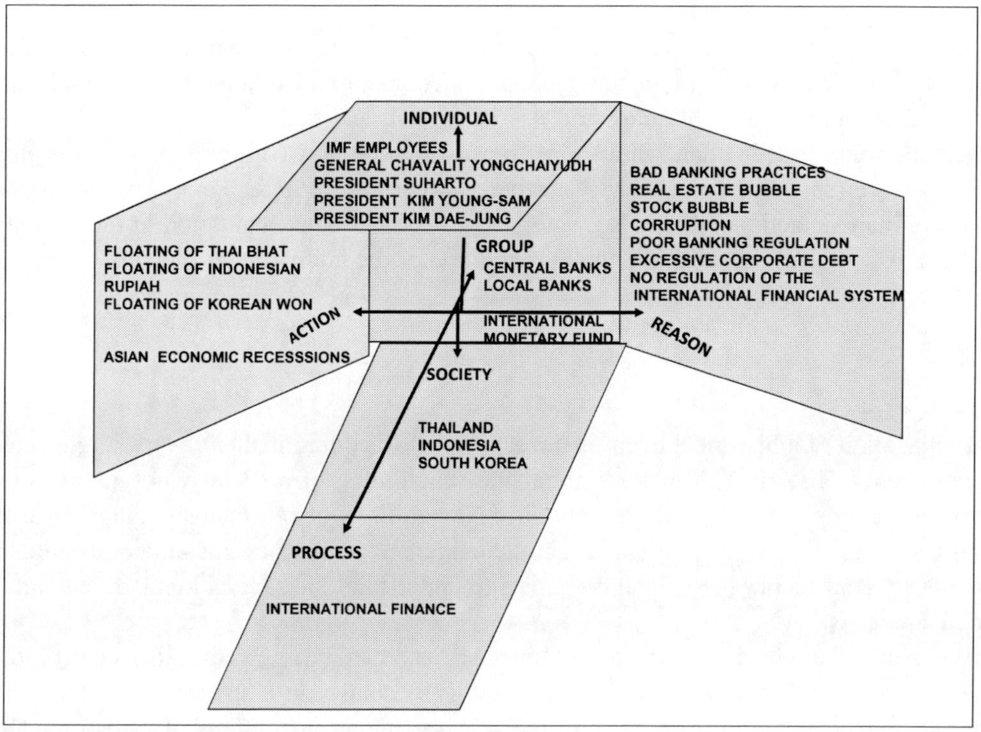

Figure 7.3. Perceptual Space of 1997 Asian Financial Crisis.

GROUPS – The groups involved in the Asian Financial Crisis were (1) local banks, local financial institutions and the central bank of each of the nations, (2) the international financial system of banks and funds in foreign countries, and (3) the International Monetary Fund (IMF). Local banks, financial institutions, and the central bank of a country constituted its national financial system. The foreign banks and financial institutions (hedge and mutual funds) constituted an international financial system. The IMF was an international organization (funded by respective government members) to help stabilize national financial systems, but not to regulate the international financial system.

PROCESS – The process was international finance, in which banks and financial institutions in foreign countries transferred capital into and out of national countries. Capital transfers occurred through short-term loans and by trading in monetary exchange rates. Stability in monetary exchange rates between countries is an essential part of stability in the international economic system.

SOCIETIES – The societies involved in the crisis were the nations of Thailand, Indonesia, South Korea (and also Philippines and Taiwan and Hong Kong).

INDIVIDUALS – The principle individuals involved were (1) leaders of the nations, General Chavalit Yongchaiyudh of Thailand, President Suharto of Indonesia, President Kim

Young-sam and President Kim Dae-jung of South Korea and (2) employees of the International Monetary Fund.

ACTIONS – The actions of the events were the floating of exchange rates (of the Thai Bhat, the Indonesian Rupiah, and The South Korean Won) and also the subsequent economic contractions in these countries, as capital flowed out of the countries, due to the decline in their monetary exchange values.

REASONS – The reasons for the withdrawal of international capital from the affected countries in the crisis were (depending upon the country) real estate bubbles, stock market bubbles, corruption in the obtaining and use of capital, and bad banking practices of giving long-term loans based upon short-term borrowed capital. Corporations in developing countries were too deeply indebted to survive a down turn in an economy. In the reasoning about the financial actions, there also was a lack of proper bank regulation in the developing countries. Also there was no international regulation of the international financial system.

Explanations

Explanation of a historical event is never simple, as seldom is there a any single cause in societal events. Historical events are about the functioning of a society, not about physical events; and a single cause-effect is only an explanation in of mechanisms, physical events. Social events usually have many factors in the explanations of its events; and one can use the cross-disciplinary theory of societal dynamics to understand the several kinds of explanations in societal events.

We recall that the issue of control in a society can be viewed a: (a) control in the managed systems of organizations in the society, (b) control in a socio-technical functional system of the society, and (c) control in the self-organizing institutions of a society. This is shown in Figure 7.4.

In the Asian financial crisis, managed systems of bank organizations occurred in local banks in a nation and in the government's central bank in the nation. Also the self-organizing sets of financial institutions divided into the set of foreign banks and mutual funds and the international banking financial aid agencies (e.g. IMF). Accordingly, in analyzing this crisis, we need to distinguish the issues of control between Local Banks and the Central Bank of a nation, between a national financial system and the international financial system, and between the self organizing set of foreign banks/funds and the international aid banks/funds.

This is sketched in Figure 7.5, in which we can summarize the many explanations involved in explaining the events of the Asian crisis, by means of:

(b) analyzing as 'managed systems' – the two kinds of banks involved in each nation, *Local Banks* and *Central Banks*;

(2) analyzing as 'self-organizing systems' – the *Foreign Financial Institutions* and the *International Financial-Aid Institutions*.

(c) analyzing the *socio-technical systems* – of each managed system and of each self-organizing system;

(d) analyzing the societal socio-technical system of the set of Local Banks and their National Central Bank, acting all-together in a nation – *National Financial System*.

(e) analyzing the societal socio-technical system of the set of Foreign Banks and International Financial Aid Agencies, acting all-together in the international scene – *International Financial System*.

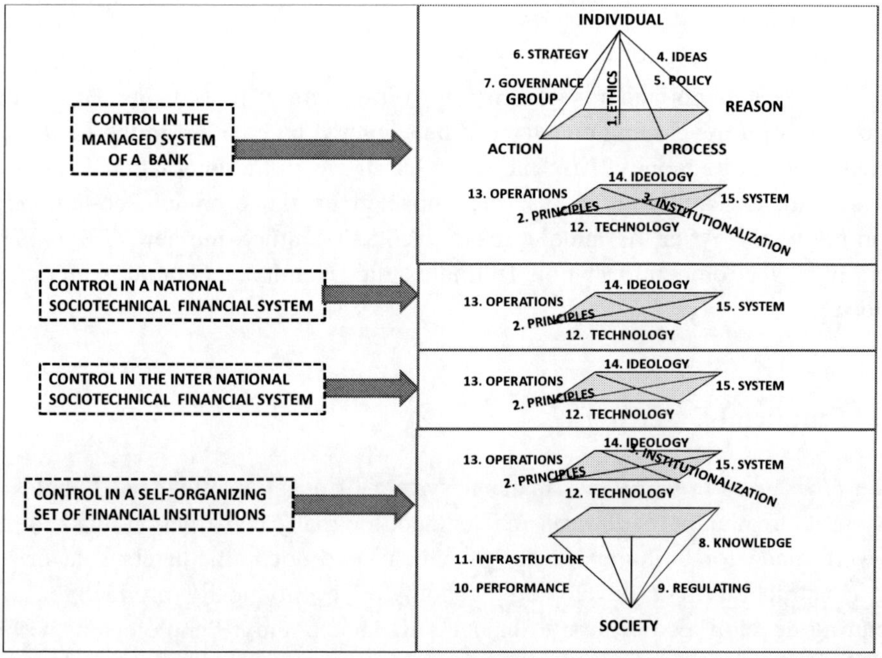

Figure 7.4. Societal Controls in 1997 Asian Financial Crisis.

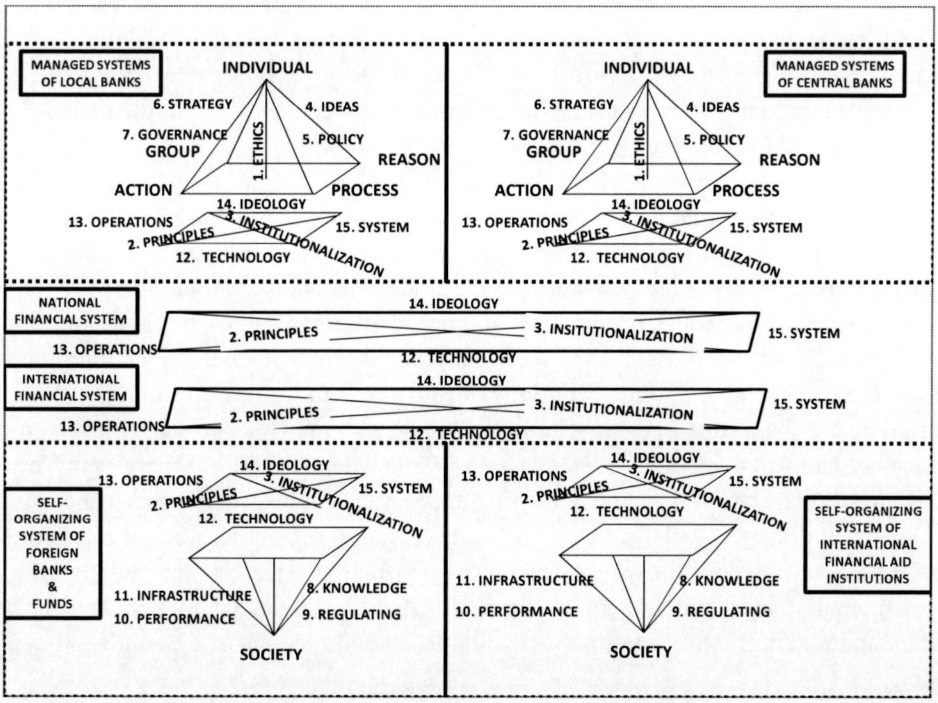

Figure 7.5. Explanations In 1997 Asian Financial Crisis.

International Financial System

Economic growth in developing countries can be disturbed by fluctuations in the *(15) System* of flow of foreign capital into and out of the country. We have seen, in the cases of the Asian Financial Crisis, that developing countries need the in-flow of foreign capital for investments in economic development – as loans to private corporations for building factories as productive investments and for government infrastructure investments. But a reversal of capital flow out of a developing country can be triggered by changes in the *(2) Principles* of the exchange rate of its money. Also the *(13) Operations* in the international socio-technical financial system, as capital flows, can be impacted by the economic conditions of each country in the world. After the middle of the twentieth century, the new *(12) Technology* of computers and electronic information facilitated the quantity and rapidity of international capital flows.

National Financial System

In the *(13) Operations* of the financial system of a nation, financial bubbles generate instability in its financial (*15) System.* In Thailand, too many construction loans (a real estate bubble) were made for buildings, many of which were not immediately sold or occupied. Building construction in a developing economy is economically productive as long as manufacturing develop also occurs, to balance off an economy. People or businesses which buy new buildings must have jobs (in other than construction) to support themselves. Building construction provides employment only once in each building, while it is being constructed. Thereafter, construction jobs must be found in building other new buildings. Manufacturing employment, in contrast, provides jobs and sustenance continuously, as long as the manufacturing business is profitable. Real estate development cannot alone sustain economic development. Other manufacturing growth is necessary to supplement it. That is why real estate financial bubbles cannot sustain any economy.

The same thing is true about financial loans in bank *(13) Operations* which foster stock market bubbles. These occur when banks make loans to stock traders to purchase stock on margin. Then the stock purchaser invests a small amount (sometimes as little as 10%) and a bank loans the rest (90%) to buy the stock. If the stock falls below that margin invested (10%) then the stock purchaser must invest more money, or the bank calls the loan and takes the margined stock as collateral. And if the stock market continues to decline, the margined stock, which the banks hold as collateral, continues to decrease in value. Too much a drop in the stock market can bankrupt a bank which has made too many large margined loans. (For example, this happened in the United State in 1929, when the highly-margined New York Stock Market collapsed, and in the following three years, banks collapsed, ushering in the Great Depression of the 1930s in the U. S.). Banks making loans to stock traders for stocks purchased on margin were following a bad banking principle. This also happened in the Asian crisis, particularly in Thailand. In national financial systems, the *(12) Technology* of computers and electronic information has facilitated the movement and quantity of capital in a country.

Also when *(14) Ideology* is capitalism but really in practice is a kind of 'crony capitalism', then government officials involved in the corruption distort a 'free market'.

(Corruption in an economy is a significant kind of 'market failure'.) In a strong economy, government regulates business activity for honesty – in order to ensure its products (goods and services) provide the public good of integrity, open competition, optimal pricing, and safety. For example in a country, the central bank should have regulatory responsibilities to license private banks and oversee that the private banks' operations are honest and follow good banking practices. In the Asian crisis, the central banks of Thailand, Indonesia, and South Korea were not properly regulating the banking practices of the private banks. Particularly in Indonesia, the central bank had allowed bribery and nepotism in financial institutions, so that such practices had a popular name: ' Korupsi, Kolusi, dan Nepotisme' (corruption, collusion and nepotism).

Managed Systems of Local Banks

In the *(13) Operations* of the Local Banks in a nation, non-performing loans may be made when the banks do not follow *(2) Practices* of good banking but instead use the bad banking practice of borrowing short-term foreign capital loans and invested them in long-term local loans. Borrowing short and lending long is one of the worst of financial principles and a certainty in leading to bankruptcy – bad banking. In the three cases of Thailand, Indonesia and South Korea, the fall in the exchange rates of their currencies trigged a financial crisis – directly because too much foreign capital had been borrowed in short-term loans but invested in long-term loans to local customers.

Managed Systems of the Central Bank

The *(13) Operations* of the Central Banks in each of the countries was to try to maintain a *(2) Principle* of fixed exchange rate to the dollar (then an international standard currency). In each nation, the central bank bought the local currency with dollars on the international exchange currency market; and it did this until the central bank ran out of dollar reserves. Then central banks had to suddenly float the national currency. This float accelerated the withdrawal of foreign capital from each nation and triggered the many failures of local banks and corporations in that nation. Although growing in financial use, still the *(12) Technology* of computers and electronic information had not been appropriately used by central banks – to properly regulate the capital activities of the local banks of a country.

Self-Organizing System of Foreign Banks and Funds

In the 1980s, the *(13) Operations* of banks and financial funds in foreign countries had begun making investments as loans to Asian banks in the Asian countries, greatly accelerating the quantity in the 1990s. The reason for this international capital flow was the *(2) Principle* of higher interest rates, than such banks and funds could find in their own country's developed economy.

Self-Organizing System of International Financial-Aid Agencies

The international financial system had grown enormously after the middle of the twentieth century, and with this was an international financial *(11) Infrastructure,* consisting of two international agencies, the International Monetary Fund and the World Bank. But neither of these had international authority for (9) *Regulating* the capital flows in the *(13) Operations* of the International Financial System.

In the self-organizing financial system of the Foreign Banks and Funds, their *(13) Operations* created large flows of capital around the world from nation to nation. Yet neither the quantity nor quality of the flows were overseen nor (9) *Regulated* by any international agency – before a crisis. But then in a crisis, the IMF tried to stop that crisis – by providing programs of financial assistance to the government in crisis. This financial intervention was always post-facto (after the fact of a crisis). It was never before a crisis (anticipating a crisis). And, moreover, during each post-facto intervention, then the IMF tried to impose financial reforms in a country's banking system. But reform imposed during a crisis, at first, only deepens the crisis. Later as the economy begins to recover, the economic reforms imposed during a crisis finally begin benefiting the economy. Reforms made before a financial crisis would be more beneficial to growing a sound economy, and avoiding a crisis.

Thus in the self-organizing international financial system, no financial institutions in any country had any of their *(13) Operations* overseen nor *(9) Regulated* by any international agency, except in crisis. Nor did the international financial institutions develop the capability of using the new *(12) Technology* of computers and electronic information to anticipate a crisis.

> *Accordingly, there had been a lack of proper (12) Technology for anticipating financial crises and a lack of proper international financial (13) Regulation about the (10) Performance (for honesty and stability) in the International Financial System.*

> *Problems in the international financial systems continued, into the twenty-first century. After the 1997 Asian financial crises, there followed the 2007-08 the Global Financial Crisis and then by the 2010-13 Euro Crisis.*

Lessons Learned

Economic development in the global world requires both an export-capable manufacturing economy and a financially-stable economy. The socio-technical financial systems involved for a financially-stable economy are several: those in the local banks and in the central banks, and in the foreign banks and funds and in the international financial infrastructure. The activities of these managed systems together form the socio-technical financial systems of a nation and of the international world. The proper integration of all these managed systems into one properly regulated, self-organizing international financial system remains a major global challenge.

There was no single cause of the Asian Financial Crisis. There were several interacting explanatory factors, and of these the most important were : (2) Principles, (9) Regulating, (11) Infrastructure, (12) Technology, (13) Operations, and (14) Ideology were most frequently discussed. From these, one can draw a lesson from this about explaining international financial crises.

In the Infrastructure of a self-organizing financial system, institutions which use bad banking Principles (i.e. borrowing short-lending-long) or crony-capitalist Ideology (e.g. corruption) and these are not outlawed by proper Regulation, then financial Operations can result in financial bubbles and in huge outflows and loss of capital around the world, followed by recessions in national economic Systems.

Information Technology has increased the rapidity and scale with which financial bubbles grow and burst and capital flows circulate in the world.

Good financial socio-technical systems are essential to economic development and economic stability and economic prosperity. Good regulation of self-organizing economic systems is essential.

Principles, Regulation, Ideology, Operations, Systems – these are the most important relations to attend for avoiding national and global financial crises.

Chapter 8

Crisis Responses – Korea and Argentina

Introduction

Global competitiveness depends not only upon the capability of a nation to export products and to import needed resources but also to survive in a world of international finance. National financial crises are triggered internally by collapsing stock markets and inflation and externally by international fiscal crises. We have reviewed the Asian financial crisis in 1997-98. But it also spread around the world to Russia and to Argentina. Russia was a special case, in the midst of a conversion from communism to capitalism. But Argentina was capitalist country, and its experience in the global financial crisis can be usefully compared to the South Korea experience. The issue we examine is the proper relation of the global financial system to a nation's economic competitiveness.

1997-99 Asian Financial Crisis: Argentina

At the end of the twentieth century, fiscal contagion extended out from Asia. We recall that the Asian Financial Crisis began in Thailand in 1996 and spread to Indonesia and to South Korea in 1997. Next in 1998, Russia defaulted on its sovereign debt; and in 2001 Argentina defaulted. South Korea quickly recovered, but Argentina did not. Why? First we will look at the Argentine situation.

Paul Blustein wrote: "The collapse of the Argentine economy, which commenced a couple of weeks after the withdrawal of the IMF mission in early December 2001, was one of the most spectacular in modern history. . . . First came the scenes of people thronging the Plaza de Mayo banging pots and pans, and mobs looting shops and sacking government buildings all over the country . . . President de la Rua was forced to resign on December 21. Then came the tragicomic spectacle of a succession of five presidents taking office over a mere ten days, ending on New Year' Day on 2002. . . . Like an engine that has seized up for lack oil, the Argentine economy ground to a virtual halt. Additional restrictions on bank withdrawals led to a breakdown in the system by which people and businesses paid each other, and the bank credit that companies needed for day-to-day commerce dried up." (Blustein, 2005)

Argentina had not always been a country on an economic precipice. Independent from Spain in 1860, it had been a rich country, with fertile land and energetic immigrants from Europe. It exported agricultural products. But the 1930s world depression hit the country's exports, and then World War II occurred. The country was divided between the few wealthy people, who owned vast agricultural estates, and the city populations with unemployment. Argentina was not a manufacturing nation. Then a new president, General Juan Domingo Peron gained popularity by employing many Argentineans in government services. Tariffs were erected against imports. The Argentine government controlled businesses through permits and license requirements and state ownership.

We can graph all this in Figure 8.1: International Commodity Prices and Production, and Export-Import Balance and Wealth Distribution and Government. In order to reduce the Unemployment Rate in the nation, the government employed people, using Sovereign Bonds to finance the Federal Deficits of the Argentine government.

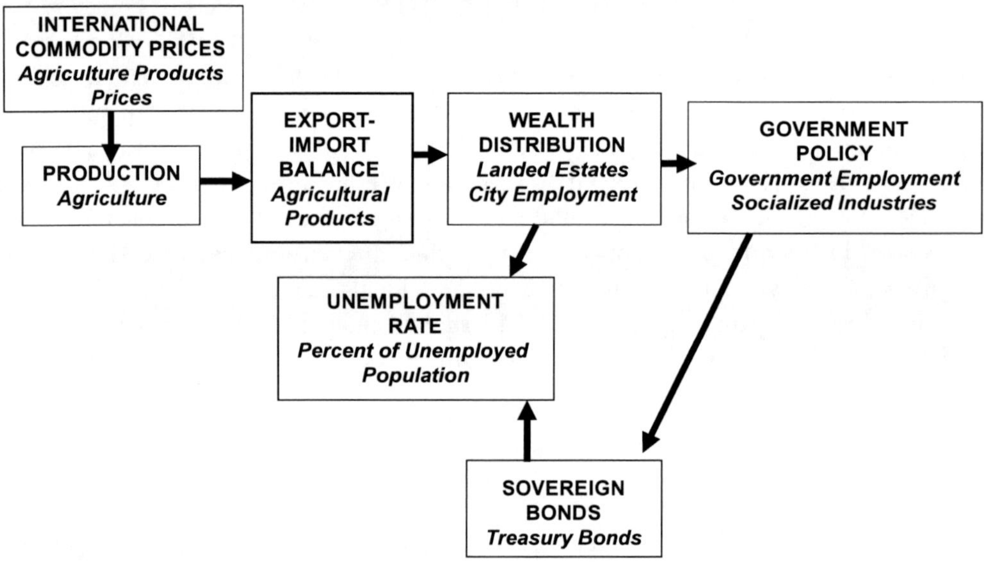

Figure 8.1. Connections to Fiscal Processes.

International Commodity Prices influence the profitability of exports of agricultural products. However because of farming technology in modern societies, agriculture does not create many jobs – primarily creating wealth for the persons (or corporations) who own the large estates. Government policies for full employment, which are predominantly agricultural, cannot depend upon the agriculture industry to provide full employment.

In 1955, the Argentine military removed Peron from office. This was the time of the 'Cold War', the war between capitalism and communism. In the Americas, Fidel Castro seized power in Cuba and became a communist dictator. Battles between the 'right wing' (capitalism) and 'left wing' (socialism) continued in Argentina, in class struggle between the wealthy and the working classes. In 1976 the military government arrested and killed many people, whom the military classified as 'left wing'. These missing persons were called the "disappeared people'' (later estimated to be about eight thousand people).

We can add this to Figure 8.2, as Government Control and Government Justice. In the ideology of a government, capitalism requires that industry be privately invested and owned. Justice of government should protect the rights of all individuals, with equal protection and law. In contrast, in the ideology of communist governments, industry is state owned and controlled and individuals have no rights.

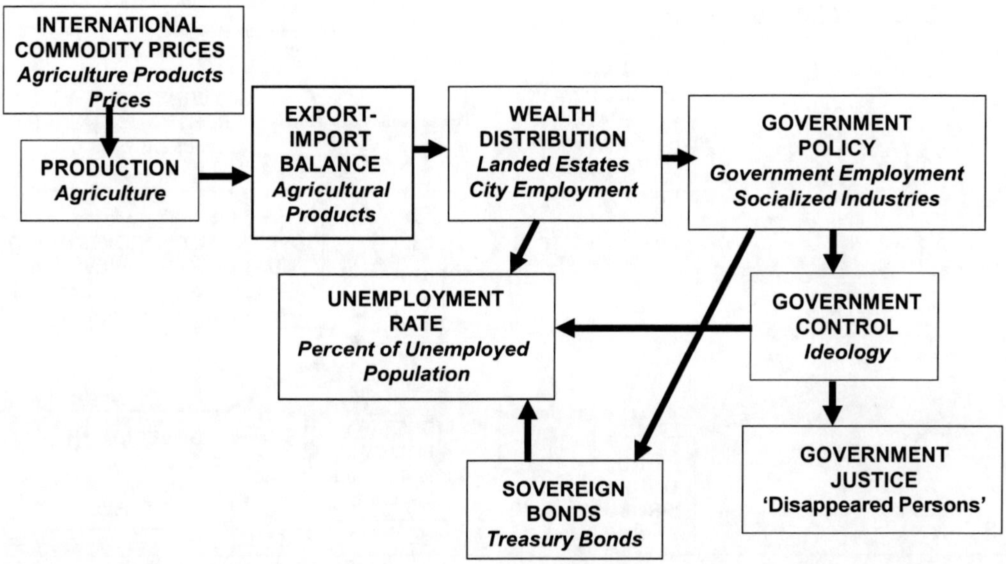

Figure 8.2. Connections to Fiscal Processes.

Under the military rule, the government ran large national deficits in funding national projects and in financing the Falklands War. The country's industries were not internationally competitive and unemployment was likely near 20% (although official figures claimed much less). In 1982, the Argentine military lost a war with Britain over control of the Falkland Islands. Discredited, the military yielded power back to a civilian rule; and Raul Alfonsin was elected president in 1983. He established a National Commission on the Disappearance of Persons. In 1985, government officials of the Military Junta were put on trial. A new currency was introduced, the 'austral' and printed. Monetary inflation jumped from 10% to 20% per month – spiraling out of control. In 1989, inflation was 200%. Riots occurred. President Alfonsin resigned his office, five months before the end of his term.

We can add all this to Figure 8.3: Government Budgets and the issuance of Sovereign Bonds impacted Monetary Policy of the nation. Also Government Revenues are the key inputs to a government budget. Taxes provide the income upon which governments operate, and corruption determines whether or not taxes are really properly collected. The disappeared persons on the chart are examples of the injustice in the government due to government ideology.

The employment rate is one measure of the state of a national economy and the stability of a government. When unemployment rate rises above 20%, an economy is in the state of a recession and if the recession last more than two years, it is in a state of depression. Governments are not stable when employment rates go above 25% (and youthful unemployment rises to 50%).

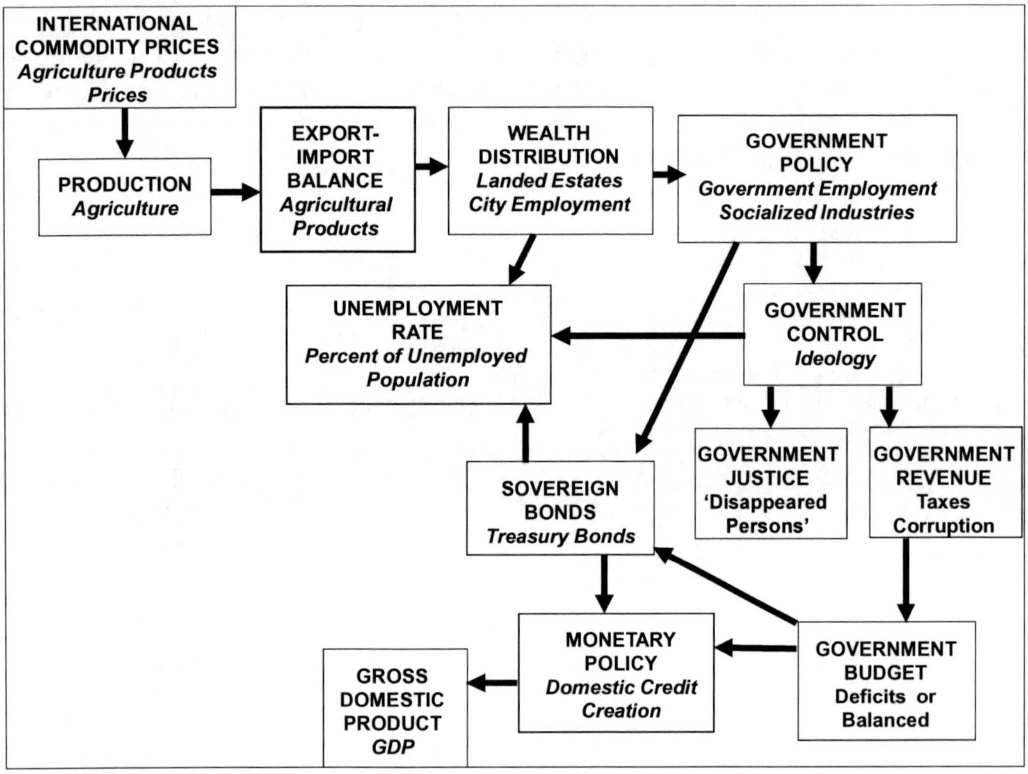

Figure 8.3. Connections to Fiscal Processes.

Governments can fund deficits by borrowing money from national and foreign investors in the form of treasury bonds. But the rate of interest of treasury bonds depends upon investor perception of the capability of the government to honor their bonds. Interest rates can rise under government policies encouraging monetary inflation. The Monetary Policy is expected to influence (control) the level of economic activity in a nation as its Gross Domestic Product (GDP). Government bonds become unsalable if the investment community anticipates a government default on their bonds. When taxes are insufficient to fund government services, then budget deficits occur. Also taxes are not collected when government officials are not diligent or honest.

In 1989, Carlos Menem, with the support of 'Peronists' won the election for president. Menem dismantled many trade barriers, reduced business regulations and began privitizations of state industries. He also established a fixed exchange rate of the Argentine peso to the U.S. dollar. In 1990, he had appointed Domingo Cavallo as Minister of the Economy, and Cavallo set the convertibility of the Austral to the Dollar at ten thousand to one. This ended the hyperinflation after 1990. The Argentine peso was reintroduced at a ratio of one to the dollar.

However, Argentina still had a large government debt, accumulated under Peron and the later military junta governments. But the fixed exchange rate of the peso to the dollar encouraged foreign financial institutions to sell Argentine bonds. And the new government continued financing the ever growing government debt with more government debt.

In July 2001, the government cut the salaries of civil servants and pensions by 13%. Strikes occurred across the nation. In August, the Government paid employees, not in money, but in IOUs. Unemployment rose to 16% (reaching 29% by December). Then the IMF refused to release the second loan payout of $1.3 billion. Bank runs began, with Argentineans withdrawing dollars from their banks. The government froze bank accounts for a year, allowing only $250 a week to be withdrawn. Protests continued in the streets, and riots occurred. On the 21st of December 2001, President Fernando de la Rua resigned, fleeing by helicopter. Riots continue throughout December 2001. The head of the Argentine Senate assumed interim presidency on December 23 but resigned on December 30. Before he resigned, the Argentine government defaulted on $132 billion dollars of its public debt (the majority of its debt).

Next on January 1, 2002, the Argentine Congress elected Eduardo Duhalde as president. On January 6, President Duhalde abandoned the peso-to-dollar convertibility. The peso exchange rate floated freely. But inflation returned, and unemployment continued to rise. The country was in the midst of a depression, throughout all of 2002 and 2003. Businesses closed. Families were out of work (and some families broke up). Thirty-five thousand people scavenged the streets, to find anything to sell.

We denote this in Figure 10.4, where the International Financial Market had anticipated a major change in the Monetary International Exchange Rate as the end of the Argentine policy of the convertibility of the peso to the dollar and stopped buying Argentine treasury bonds. A 'Central Bank Reserve" is the amount of internationally-recognized money which a government national bank holds as an international-standard monetary reserve (e.g. dollars or gold or euros). When the International Financial Market stopped buying Argentine bonds, this created the Bank Runs on Argentine banks, which in turn triggered an Economic Depression, which in turn drastically drove up the Unemployment Rate.

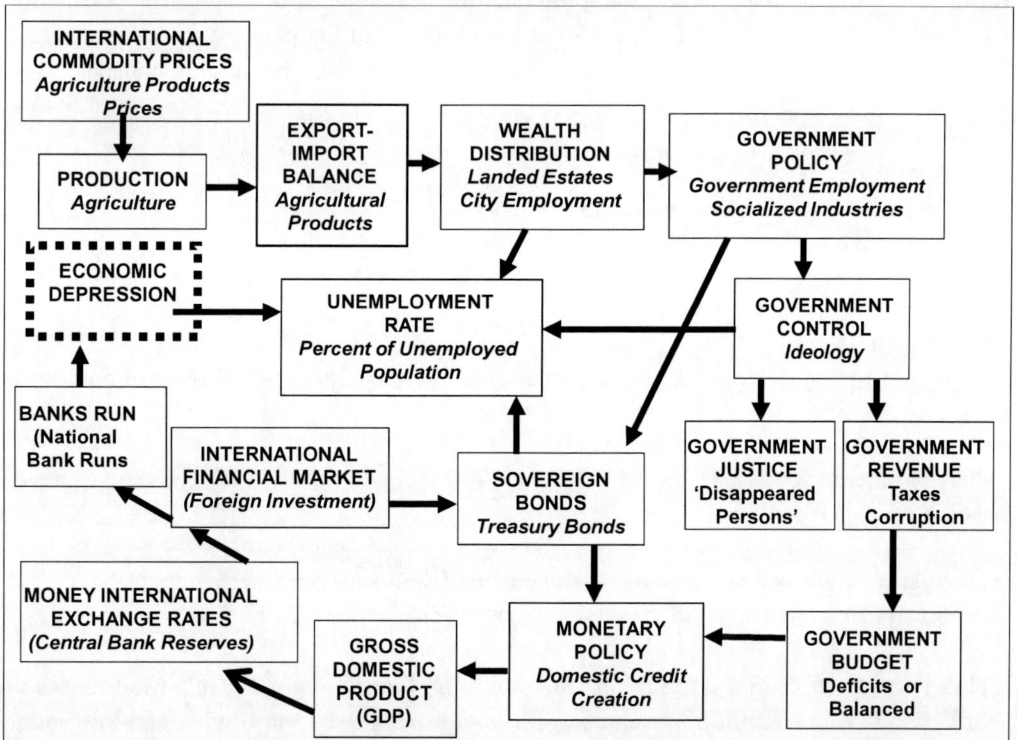

Figure 8.4. Connections to Fiscal Processes.

As sketched in the diagram of Figure 8.4, we see that this is a complicated story; but it is a depiction of the micro-economic structure of Argentina, in the decade from 1991 to 2001. It is in such detail of the micro-economics of a nation that one must go to in order to understand the economic history of such a period. *Topological graph models can depict the detailed complications of the micro-economic structure of a society in a given period.*

What we next review is that, when the International Monetary Fund (IMF) tried to help out Argentina, its policies were not informed by such micro-economic detail. Instead the IMF officials based their policies upon a macro-economic theory which did not match the empirical context of Argentina. The IMF policies did not help in the Argentine crisis but only exacerbated it.

IMF Policy Model

The International Monetary Fund (IMF) negotiated with respective countries in the crisis to provide loans, intending to avoid financial collapses. But in each case, IMF conditions and loans to governments did not prevent financial collapse. Why not? To understand this, we review the fiscal model which IMF officials used to guide their policy decisions – their macro-economic model – called, by the IMF, the 'Polak Model'. Ngaire Woods wrote: "Early on in the life of the IMF, a particular model emerged . . . . the Polak model. It offered staff a way to diagnose and prescribe conditions for any economy facing a balance of payments crisis. . . The original model required few data. It focused attention on a key variable that governments could control – domestic credit creation."(Woods, 2006)

In 1989 and 1998, J. J. Polak published his model, and his paper was reprinted later as collected essays. (Polak, 1989) (Polak, 1998) (Polak, 2005) The model consists of four equations:

(1) $\mathbf{I = kP}$

Imports (**I**) are proportional (**k**) to GNP (**P**).

(2) $\mathbf{E - I = (\Delta R - C)}$

Exports (**E**) minus Imports (**I**) are equal to (Change-in-Reserves (**ΔR**) minus Net-Capital-Inflow (**C**) of the non-banking sector).

(3) $\mathbf{\Delta M = k\,\Delta P}$

Change-in-Money-Supply (**ΔM**) is proportional (**k**) to Change-in-Gross-National-Product (**ΔP**).

(4) $\mathbf{\Delta R = (\Delta M - \Delta D)}$

Change-in-Reserves (**ΔR**) equals (Change-in-Money-Supply (**ΔM**) minus Change-in-Domestic-Credit (**ΔD**).

The first equation is about imports (I) and exports (E) in a nation, asserting that:
Imports (I) are proportional to the total revenue (P) in a nation.

This is a questionable assumption. In a nation, what-is-imported and what-is-exported depends upon what-cannot-be-produced-in-a-nation *(Imports)* and what-can-be-produced *(Exports)*. But what can be produced and what needs to be imported bears no relation to the total revenue. Instead, what-can-be-produced depends upon the industrial technology in a nation and the resources of a nation. In 1998, Argentina could not produce automobiles, but South Korea could. Argentina, with its vast grassy plains, could produce and export agricultural products; South Korea with limited resources for rice production could just feed its own people. In this way, macro-economics (big picture) depends upon micro-economics (detailed structure). Each country's economic structure (production, banks, companies, real estate, government budgets) was different and unique.

Argentina exported agricultural products (e.g. frozen beef), while South Korea exported manufactured products (e.g. semiconductor chips and automobiles). Argentina imported manufactured products, and South Korea imported oil and iron ore. Thus to keep Argentina running, employment in the agricultural products was vital to maintain. While to keep South Korea running, imports of oil and iron ore were vital. The macro-economic policies of IMF needed to differ for the two countries at the time, because of their different micro-economic contexts.

So the first equation **I = kP** was not true. Imports (I) in a country was not directly related to the Gross National Product (P). More GNP could provide more money to purchase imports. But, as we saw in the different countries, the level of imports was heavily financed by foreign credit (by borrowing from international credit markets). It was excessive borrowing of money (foreign loans) which increased the level of national imports, far more than their GNP did. Polak's first macro-economic assumption about imports and GNP was not historically correct in Argentina in the Asian financial crisis.

The second equation was also about imports (I) and exports (E): **E - I = (ΔR – C).** It assumed that the difference between Exports (E) and Imports (I) depended upon changes-in-money-reserves (ΔR) in the central bank of the nation and upon the inflow of capital (C) into the non-banking sectors. Exports-Imports (E-I) increase as an increase in the national bank reserves and as a decrease in capital flows. Did in fact exports increase over imports when the central bank reserve of money increased in Argentina? No. In Argentina, exports depended upon the international price of beef products; when the price fell, Argentina exported more frozen beef, when the price rose, Argentina exported less. In Argentina, the difference of exports-to-imports bore little relation to the amount of gold held by the Argentine central bank. What the central reserves were related to was the exchange rate of the Argentine peso.

Also did in fact exports increase over imports when the capital flows to manufacturing investments decreased? No. Argentine commerce had not invested in manufacturing but in agriculture. But in South Korea, the opposite was true. Exports of manufactured goods increased when capital flows increased as investments in manufacturing.

The second equation **E - I = (ΔR – C)** assumed that the difference between Exports (E) and Imports (I) depended upon changes-in-money-reserves (ΔR) in the central bank of the nation and upon the inflow of capital (C) into the non-banking sectors. This meant, for example, if a nation buys gold (or dollars) and companies in the nation borrow less in foreign loans, then exports will increase over imports. This assumption is about how a financially stronger country (with more gold and less foreign loans) could increase exports (or decrease imports). Certainly financial strength in a nation could increase exports – but only if that increase in reserves (i.e. gold) is used to invest in more manufacturing capability in the nation. Thus Polak's second macro-economic assumption is contextually dependent on how a country uses its wealth, upon is micro-economic structure.

For example, in the case of South Korea, inflows of foreign capital were financing manufacturing investments. Yet the decrease in capital flow did not increase exports but paralyzed the manufacturing sector bringing about the economic collapse. In the case of Argentina, foreign capital inflows were going to finance government services and not to private manufacturing investments. So in both countries, a dependence of the difference of exports and imports had *no direct relationship* to increased central bank reserves and decreased foreign loans. This macro-economic assumption was not historically true in these two countries in which IMF used it as policy.

For the third equation **ΔM = k ΔP,** the assumption was that Change-in-Money-Supply (**ΔM**) is proportional (**k**) to Change-in-Gross-National-Product (**ΔP**). In Argentina, this was periodically not true. Governments inflated money, increasing the supply of money, without increased the GNP. Inflation impacted not the gross revenue of Argentina, but Argentina's exchange rate. Polak's equation does not distinguish between a real increase in monetary supply (productive expansion of money and credit in a national financial system) and an artificial increase through inflation. Obviously, Polak meant **ΔM** to be measured as real increase. And a real increase in money-credit can help increase GNP when the increase in money is invested in increasing manufacturing and services. But this also is contextually dependent, the micro-economics of how a nation uses its change in money to invest or not invest in expanding production. Polak's third equation was not true in Argentina, because the government inflated Argentine money by borrowing foreign money to finance huge and continuing government debt.

For the fourth equation **ΔR = (ΔM - ΔD),** Change-Reserves (ΔR) did not actually equal the Change-in-Money-Supply (ΔM) minus Change-in-Domestic-Credit (ΔD). Domestic Credit froze up as banks stopped lending because of their insolvency. The central bank reserves did not increase at all; and instead the economy stopped. In all the countries, the empirical reality about Reserves of dollars (R) was quite different from Polak's assumption. Changes in reserves (ΔR) depended upon the exchange-rate-of-the-local-currency. Reserves were used by a central bank to defend a national fiscal policy of stability of local currency. In the earlier phases of the Asian crisis, the Thailand central bank had purchased baht with dollars to defend baht exchange rate. The Indonesian central bank had purchased rupiah with dollars to defend the rupiah exchange rate. The Korean central bank had purchased won with dollars to defend the won exchange rate. The Argentine central bank bought pesos with dollars to defend the peso exchange rate. Contrary to Polak's assumption, it was an exchange-rate fiscal policy which depleted national bank Reserves (not the changes in money-credit supply).

Ngaire Woods explained: "Crucially, it (the Polak model) linked a country's domestic economic policies to its balance of payments position. This opened the door for IMF conditionality (conditions for making an IMF loan to a country). . . . The starting point of the Polak model is what was known (in macro-economic theory) as the 'absorption approach' to the balance of payments – that a country with a balance of payments deficit was absorbing too many resources in consumption and investment, relative to what that country can produce. With a couple of simplifying assumptions, it will follow that a country which increases domestic credit too rapidly will encounter increasing balance of payments deficits reflected in a loss in central bank reserves. . . On the basis of this analysis, when a country has a balance of payments deficit, the Fund's prescription focuses on reducing government spending, increasing taxes, and reducing domestic credit creation." (Woods, 2006)

The application of a uniform macro-policy by the IMF that ignored national micro-economic structural details did not help any of these countries – but only deepened each crisis.

Empirically, any macro-economic model of an economy is contextually dependent upon the detailed micro-economic structure of that economy.

Why did Polak make these assumptions? One can only suppose that Polak's economic thinking traced by to traditional economic thinking, which had been earlier developed in 1800s in the industrial revolution in Europe (see, for example, Polyani, 1944). Then economic thinking had been focused upon exports, because what industrial technology was doing then was producing too much product to be consumed locally. Industrialization is a business phenomenon which requires export of the large amount of product which industrial technology can produce in the factories.

Then English economic thinking had properly focused upon exports and imports. More exports – good! More imports – bad! A positive export-minus-import brought economic prosperity to the wealth in England; and a negative export-minus-import brought economic recession in England. Economic recession was bad because, capitalists lost wealth and labor was unemployed.

But economically what could be done to increase exports and decrease imports? The economic thinking then was to increase unemployment – as a way to reduce labor prices thus reducing production costs and thus decreasing export products prices – thereby increasing exports and profits. This was the solution, control over employment. Firing labor increased exports; and when exports increased, labor could be rehired. A capital solution!

Looking back historically, there was, in the 1800s, a micro-economic basis for the validity of this kind of economic thought – in the industrialization of Europe. Then manufacturing automation was not yet a technology; it would occur later in the 1900s. In the 1800s, manufacturing machinery was the new technology. So back then, manufacturing machinery needed many people to run the machines, as industrial labor. Economically in the English industrial revolution, labor was a large part of production costs – roughly 50% of production costs was in labor and 50% in materials. (Then the costs-of-production had not included the cost of factory investment, because manufacturing machinery was directly purchased and not financed by loans). Accordingly, since material costs could not be controlled by English capital, labor costs could, by increasing unemployment and making people more willing to work at lower wages.

But in the modern twenty-first century, labor costs in most manufacturing are no longer large costs. Labor costs are now around only 5% of direct production costs; while material costs are still high, around 35%, and overhead consists of the remainder of the 60% of the direct costs. (The exception to this is the cost of labor in clothing assembly, as still labor costs there have been at least 35%, with material costs 35% and overhead 30%.) The difference in manufacturing costs between the nineteenth century and the twenty-first century has been in the technologies of manufacturing automation – which had rapidly progressed all through the twentieth century. For this reason economic thinking about an industrialized economy in the 1800s has not been appropriate for modern economic thinking about present industrialization. The Polak model constituted economic thinking by IMF officials, which was 100 years out of date about manufacturing technology.

The Polak model was aimed at increasing a national trade balance (increasing the ratio of exports/imports) – all dependent upon a causal connection between export-imports and unemployment (unemployment created by reducing business through increasing taxes and reducing credit). But this was not true in Argentina. Argentine unemployment produced depression and government instability, but did not increase exports. The Polak model missed important parts of Argentine reality.

Also the IMF Polak model assumed that 'Monetary Policy' (which is the control of domestic credit creation) will directly influence 'Gross Domestic Product' (which is the total level of revenue generated annually in a national economy). This is an important assumption, which may or not be true. We saw this was not true of Argentina, as neither inflation nor deflation increased GDP. Argentine GDP was more influenced by the international commodity prices, since Argentina was an agricultural exporting nation. This was why, when applied to Argentina, the IMF Polak model resulted in bank runs and economic recessions. And this also had occurred earlier in Thailand, Indonesia, and South Korea.

1. The IMF Polak policy model was seriously inaccurate in modeling economic reality.
2. The Polak assumption (that raising the unemployment rate directly stimulates increased exports) was seriously incorrect – in explaining economic reality.

To summarize, economic thinking that assumes laying-workers-off-to-reduce-production-costs-will-turn-around-an-economy is out of date. Unemployment may or may not reduce product prices, depending upon how major a factor is labor cost in production. In the twenty-first century, the major factor in the manufacturing costs is no longer labor wages but investments in automation. Most of the wage costs in industry now is in management salaries and bonuses and not in labor. Over the last century, automation has increasingly taken labor out of production effort and cost. Thus exports may or may not be increased, as labor wages are lowered by unemployment. Now there is no direct connection between increasing exports and increasing unemployment – not in Thailand, Indonesia, South Korea, nor Argentina. The basic assumption, in the traditional economic thought underlying the IMF Polak model, is no longer historically appropriate – that increased unemployment will increase exports.

Nagaire Woods wrote: "In defining its craft, the IMF is heavily constrained both by its capacity and by the limits put upon it by its most powerful members. Within these constraints for a long time, the Polak model (and successive financial programming models) made life relatively easy for the Fund. . . . That said, financial programming was severely challenged during the 1980s as the IMF sought an appropriate response to the debt crises that afflicted so many developing countries. . . The downside of professionalism for the IMF (and the World Bank) is that there is very little room for local knowledge. Local knowledge is messy, political, intractable, and very difficult to make judgments about." (Woods, 2006) But also 'local knowledge' provides the empirical detail about whether or not an abstract model really captures reality – or misses it. Fiscal policy which is not modeled upon reality can do a great deal of social harm. Argentine citizens paid for the corruption of Argentine officials and for the incompetence of IMF professionals. Models, upon which policy is based, should capture societal reality – completely and correctly.

The Polak model, which sounded nice when expressed in math symbols, was simply incorrect when expressed in historical reality.

Economic models are unreal, when based upon economic theory which is not empirically grounded in economic history.

Societal harm is done by the use of unreal policy models based upon flawed societal theory.

These conclusions have become widespread among some contemporary economists. We have quoted from Ngaire Wood's depiction of IMF policy in Argentina. Others, such as Joseph E. Stiglitz, have also re-examined IMF policies of the time: "Perhaps of all the IMF's blunders, it is the mistakes in sequencing and pacing, and the failure to be sensitive to the broader social context, that have receive the most attention." (Stiglitz, 2002)

These mistakes derive from an incomplete theory about society (by some economists) that a society is mostly an economic system. Stiglitz wrote that the 'mistakes' of IMF consisted of: " . . . forcing (economic) liberalization before (societal) safety nets were put in place, before there was an adequate regulatory framework, before the countries could withstand the adverse consequences of the sudden changes in market sentiment that are part and parcel of modern capitalism; forcing policies that led to job destruction before the essentials for job creations were put in place; forcing privatization before there were adequate competition and regulatory frameworks." (Stigliz, 2002)

This is an interesting list of necessary societal 'priors' (before's) for modern economic systems to properly function:

– adequate regulatory framework,
– adequate competition,
– safety nets for economic liberalization,
– policies for job creation rather than destruction.

But why were these societal-prior-conditions not recognized in the earlier Polak policy model, upon which IMF officials had relied? Stiglitz wrote: "Many of the sequencing mistakes reflected fundamental misunderstandings of both economic and political processes, misunderstandings that were particularly associated with those who believed in market fundamentalism. They argued . . . that once private property rights were established, all else would follow naturally – including the institutions and the kinds of legal structures that make market economies work." (Stiglitz, 2002)

The reason for this is that in societal systems, nothing is simply 'natural' as in the nature of physics – where everything is naturally mechanistic, naturally physical mechanisms. Instead, social things in society are not mechanistic but functional. The 'nature' in society is 'function' – functional activities by individuals, by groups of people, by businesses, by governments. Societies are functional things and require many functional aspects, such as: leadership, integrity, justice, capability, ethics, power, technology. A properly functioning economy in a society requires a lot of societal-prior-conditions.

Why has this basic understanding about society not been pervasive in economics? Stiglitz wrote: "Behind the free market ideology, there is a model, often attributed to Adam Smith, which argues that market forces, the profit motive, drive the economy to efficient outcomes as if by an *invisible hand.* One of the great achievements of modern economics is to show the sense in which, and the conditions under which, Smith's conclusion is correct. It turns out that these conditions are highly restrictive. . . . Significantly, there are desirable government interventions which . . . can improve upon the efficiency of the market." (Stiglitz, 2002)

It is these 'desirable government interventions' which IMF policy got wrong. IMF did not push for the government: (1) to establish the proper 'safety nets for economic liberalization', (2) to establish an 'adequate regulatory framework', (3) to establish policies for job creation rather than destruction, and (4) to regulate commercial and financial institutions to achieve 'adequate competition'.

Stiglitz wrote: "If the crises had a familiar pattern, so too did the IMF's responses. If provided huge amounts of money . . . so that the countries could sustain the exchange rate. . . It was thus, in part, a bailout to the international banks as much as it was a bailout to the country. The lenders did not have to face the full consequences of having made bad loans.." (Stiglitz, 2002)

The economic justification of 'bailout to the international banks' was in the IMF conditions to 'reform' a country. Stiglitz wrote: "The IMF combined the money with conditions. . . . higher interest rates . . cut backs in government spending and increases in taxes. The IMF would claim that imposing these conditions was the responsible thing to do. . . The IMF programs, with all of their conditions and with all of their money, failed. . . . As a crisis progressed, unemployment soared, GDP plummeted, banks closed. . . The crisis economies . . were clearly threatened with a major downturn and needed stimulation. The IMF programs pushed exactly the opposite course. . . . Today, the IMF admits that the fiscal policy it recommended was excessively austere. " (Stiglitz, 2002)

Thus the problem of correcting bad fiscal policies (which lead to too much government borrowing) does not lie in IMF policies of more loans and restrictive reforms during a crisis. Instead it lies with how a government handles a crisis and fixes its own financial and commercial system. We next examine this difference between Argentina's fiscal crisis and South Korea's fiscal crisis. South Korea fixed itself. Argentina did not fix itself. The IMF fixed nobody.

Societal prior-conditions for an effective economic system require that each society fix its own functioning – in a proper and patriotic partnership between government and commerce – within a capitalistic economy.

COMPARING SOUTH KOREA AND AGENTINA ECONOMIC CRISIS MANAGEMENT

We examine how the South Korean government and industry dealt with the crisis (fixed their prior problems existing in the Korean economy). South Korean reform set the stage of a rapid resurgence of economic growth. A decade later, South Korea was one of the leading economic nations in the world. While the Argentine economy eventually recovered, it did not grow as successfully. In 2013, South Korean was a globally competitive industrialized nation, while Argentina was still primarily an agricultural-exporting nation.

Yoon Je Cho summarized the Korean crisis: "The Korean economy faced a severe financial and currency crisis in November 1997 and started an IMF program in early December 1997. It fell into a deep recession, with real GDP contracting 67 % in 1998. The economy started a rapid recovery in late 1998, with growth rates of 10.7 % in 1999 and 8.8 % in 2000." (Cho, 2002) The economic decline was quick. But surprisingly, the economic recovery was quick. Rapid economic decline in a financial crisis is a standard pattern, but rapid economic recovery is not standard. To understand the uniqueness of the Korean recovery, we examine how the Korean partnership of government and industry set to work immediately to fix the crisis.

Kyu-sung Lee wrote: "In December 1997, domestic demand continued to contract sharply (and continued to contract during the first quarter of 1998). Activity in the construction sector spiraled downward, and imports of machinery and orders for domestically produced machinery slowed drastically. Wholesale and retail sales fell. Sales of consumer durables . . .dropped. . . . Production was slumping across the board. . . At the same time, consumer prices surged and put enormous pressure on households. . . Housing prices declined, and the real estate market was in a severe downturn. Inflation was fueled mostly by spiraling import prices because of the weakening won. . . . The key determinant of inflation was the won/dollar exchange rate. . . . With corporate bankruptcies on the rise, unemployment surged beginning in December 1997 – from 570,000 in November 1997 to 1,200,000 in February 1998." (Lee, 2011, location 3711)

Kyu-sung Lee was the Minister for Finance and Industry in D. J. Kim's government in 1988, and he participated in the formulation of government policies to contain the crisis and reform the financial and commercial infrastructure of South Korea. Central to properly handling the crisis was not IMF macro-economic policies but Korean government micro-economic structural policies.

Korean government policy focused upon both the macro- and micro-economics of Korea. Yoon-Je Cho wrote: "The initial crisis resolution strategy under the IMF program with Korea was composed largely of two parts: macroeconomic policies and structural adjustment measures (microeconomics). . . . The main goal of the (macroeconomic) policies was to stabilize the exchange market and to improve the current account. The main components of the structural reform policies were financial restructuring and the rapid adoption of global standards in financial supervision, accounting standards, disclosure requirements, and corporate governance. The main goal was to increase transparency and accountability in management and thus to improve economic efficiency." (Cho, 2002)

How was this done? It was begun by the Korean government under the new President D.J Kim, eliciting commercial cooperation by private firms in the reform of Korean economics. Kyu-sung Lee wrote about the five principles the President Kim imposed upon chaebols in order to gain government assistance in the Korean financial crisis of 1997-99: "In principle, corporate restructuring had to be undertaken by companies voluntarily. Financial companies were to enhance their business practices through strict evaluations. The theme running through the five principles of business restructuring was that companies must be managed on the basis of transparent and accountable practices in corporate governance, financial soundness and competitive business structure. The five principles were:

> "First, business transparency had to be drastically enhanced. . . . "
>
> "Second the cross-debt guarantees that chaebol-affiliated companies provided to each other had to cease, so as to ensure the financial independence of the companies in business groups and prevent one failing company from dragging down an entire business group . . . .

"Third, the fundamental financial structure of companies had to improve.

"Fourth, companies had to choose a core competence and strengthen their business ties with SMEs.

"Fifth, corporate management and dominant shareholders had to be had strictly accountable for their actions." (Lee, Kyu-sung, 2011)

Now we can compare these Korean government's reform principles (in Lee's words) to Stiglitz's list (of necessary societal-prior-conditions for modern economic systems to properly function) – comparing empiricism to theory:

Stiglitz – adequate regulatory framework

The Korean principle was: "First, business transparency had to be drastically enhanced. "Second the cross-debt guarantees that chaebol-affiliated companies provided to each other had to cease, so as to ensure the financial independence of the companies in business groups and prevent one failing company from dragging down an entire business group . . . . "Third, the fundamental financial structure of companies had to improve."

Stiglitz – adequate competition

The Korean principle was: "Fourth, companies had to choose a core competence and strengthen their business ties with SMEs. Fifth, corporate management and dominant shareholders had to be had strictly accountable for their actions."

Stiglitz - safety nets for economic liberalization

This was not addressed initially in the Korean approach.

Stiglitz - policies for job creation rather than destruction.

This was not addressed initially in the Korean approach.

The Korean government principles in the crisis were aimed at two kinds of problems: *macroeconomics* to get the money and credit flowing again, and *microeconomics* to reform the structures of the financial and commercial institutions (for competence, accountability, effectiveness, transparency, competitive). (We note that the five principles did not initially include principles about safety nets and job creation; but these were also to be later addressed.)

The interesting thing here is the connection between macroeconomics (government fiscal policy) and microeconomics of Korean infrastructure (businesses in an economy).

Unless the microeconomic infrastructure in an economy (its businesses) is – competent and accountable and effective and transparent and competitive – the macroeconomic government fiscal policies cannot work.

Unless businesses pay taxes, governments cannot have balanced budgets to fund government services.

The government set the pace for the corporate restructuring. Lee wrote: ". . . the government decreed that review of the viability . . . of large companies and SMEs would have to be completed by the end of June (1998). . . . And 210 financial institutions . . were directed to conclude restructuring agreements (with debtor companies) by June 25. . . On June 18, creditor banks announced a list of 55 companies they judged to be nonviable (to be closed) along with a list of corporate debtors they considered distressed. . . . (Also) five banks were closed." (Lee, 2011)

The government reviewed 210 of the Korean financial institutions (banks and others), with the review to be finished in six months (January to June in 1988). In June, the review recommended that five banks were to be closed and 55 banks to be restructured.

But this announcement of bank closures set off another wave of fiscal crisis in June: "When the announcement of the closure of five failing banks was made on June 29, the business operations of the banks went into complete paralysis. . ." (Lee, 2011)

During the crisis times and at the same time trying to reform an economy, the beginning of reforms themselves can trigger more crises. Lee wrote: "The government had again to act to keep the banking system operating. . . The government ordered full cash payments for checks issued by the failing banks and the exemption of the bills and debt instruments of the customers of the banks from delinquency. . . For seven banks that received conditional approval for turnaround, restructuring proceeded. . . . Also twelve banks, with capital ratios exceeding 8 percent, were allowed to merge (if they wished). The restructuring (of merchant banks and insurance companies) . . . was to be led primarily by the shareholders. . . . (But) the business licenses of fourteen merchant banks were revoked. The restructuring of insurance companies began . . Seven life insurance companies and two non-life insurance companies were order to submit turnaround plans within a month. . . On August 11, the FSC suspended the business of four life insurance companies. . . . (Lee, 2001, location 4827)

Ultimately, five commercial banks, fourteen merchant banks, and seven life insurance companies were closed. The others were ordered to restructure their finances to a sound fiscal basis. This was the effect of the first micro-economic reforms of Korean government policy. The restructuring by the government focused first on the financial sector, and this began to produce economic benefits. Lee wrote: "A first benefit . . was the positive effect on the turnaround efforts of financial companies. . . Second, as the soundness of the banking sector improved, the financial sector found some breathing room from the stifling credit crunch . . . Third, . . restructuring send a powerful message to foreign investors and analysts who had been skeptical of Korea.." (Lee, 2011, location 4867)

Along with reforming the infrastructure of the financial sector, reform of the industrial sector was simultaneously proceeding. Lee wrote: "The closure of the 55 nonviable companies was completed in June, and the government shifted the focus of restructuring to helping companies turn around their businesses. . In the restructuring, the top five chaebols . . . were reviewed on their agreements to improvement in financial soundness, . . . For the other chaebols, the plan was to initiate, along with creditor banks, workouts aimed at turnarounds. (Lee, 2011) The companies had overextended their credit. Some chaebols held debt-to-equity ratios of 400-to-1. Normal global companies vary around 100-to-1. The chaebols agreed to reduce their debt load down to 200-to-1. The top five chaebols had sufficient access to funding to straighten up their finances. The remaining chaebols had to work with their banks to 'work out' (reduce) their debt. Fifty-five companies were closed.

But also the government had to attend to the problem of unemployment *(Stiglitz– policies for job creation rather than destruction).* About this, Kyu-sung Lee wrote: "To minimize the loss of existing jobs, the government sought to boost the credit guarantees and funding for SMEs and thus help them restore business stability." (Lee, 2010)

The rapid 'turn-around' of the South Korean economy from the Asian crisis occurred for two reasons. First, Korean industry and government had been building a strong export economy since 1963. This strong industrial basis could be reformed quickly, by strengthening existing micro-economic infrastructure. Second, the Korean government formulated comprehensive and active policy to encourage the infrastructure reforms.

We analyze the Korean economic reform event in the format of a societal perceptual space, Figure 8.5.

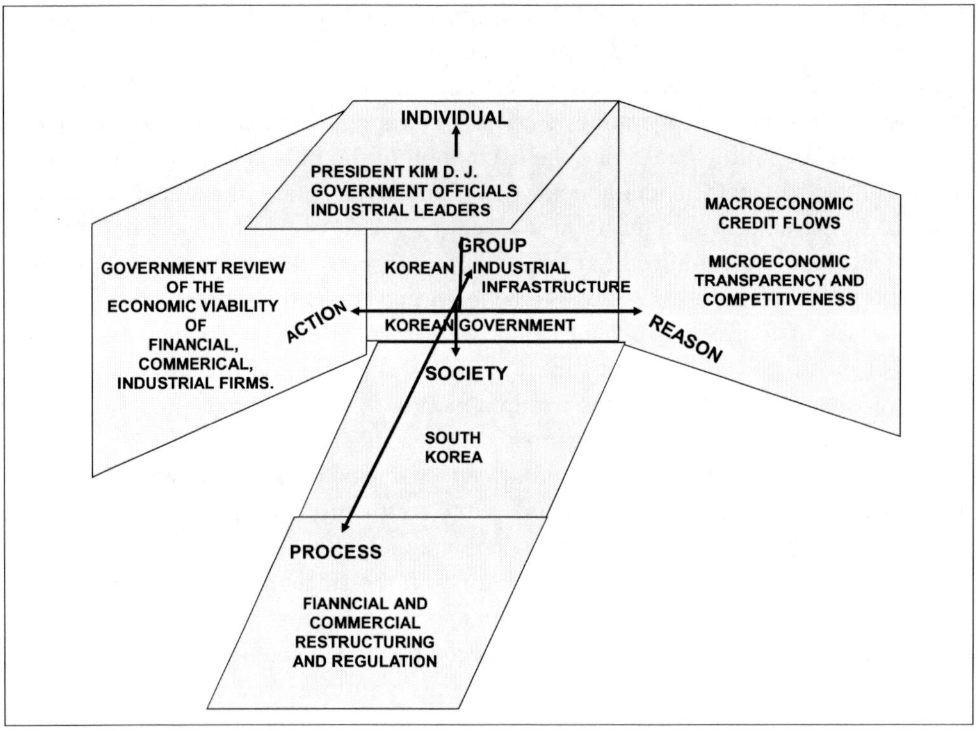

Figure 8.5. Perceptual Space of South Korean Financial Reforms in 1998.

ACTION – The action in the event was an immediate government review of the economic viability of Korean firms in the financial sector (banks, insurance firms) and the commercial and industrial sectors (chaebol groups and SME businesses). The review determined which groups needed to improve their fiscal and governance operations.

REASON – The reasoning underlying government policy and commercial reform centered upon (1) macro-economic concerns of maintain credit flow and monetary stability in the Korean economy and (2) micro-economic concerns of firms in the economy operating with fiscal transparency and competitive effectiveness.

PROCESS – The process was government agencies review of each firm in the Korean economic infrastructure, grouping these firms into classes of relative fiscal performance and imposing regulations upon performance and governance to restructure the firms into internationally competitive and responsible economic agents.

GROUPS – The involved were government agencies and Korean firms.

SOCIETY – The Republic of Korea was the society in which government and industry cooperated together to stabilize economic activity and continue national growth.

INDIVIDUALS – Individuals involved in the events were both government officials, led by the President Kim D.J,. and private individuals as heads of the financial and industrial firms of South Korea.

Korea recovered rapidly because it had been building a manufacturing exporting economy; and a government-industry partnership acted together to re-regulate financial and commercial toward financial soundness.

ARGENTINE RECOVERY

Argentine economic recovery was not so swift. Finally in 2004 and 2005, the Argentine economy began to recover, as cheap pesos made Argentine agricultural exports inexpensive. Edward Glaeser wrote: "In its pre-World War I heyday, Argentina thrived as a trading giant shipping beef and grain abroad. After World War II, formerly poor countries including Japan, Korea and Italy followed an export-led model to wealth. . . . But Argentina did turn inward. Peronism was not only protectionist, but it also favored large state enterprises and significant regulation of the economy. Neither strategy had been particularly good for growth. Argentina's inbred banking system historically had trouble weathering severe shocks. Decades of political instability made property rights insecure and investment unattractive. Argentina was cursed with bad policies that bear much of the blame for the country's problems. . . . The story of Buenos Aires is . . . a city surrounded by a vast, fertile hinterland. Buenos Aires grew as a center for transporting agricultural products east. The frigorificos, refrigerated ships, greatly increased its ability to ship beef. Clothing was also Buenos Aires's largest industry. " (Glaeser, 2009)

Ian Mount wrote: " Argentina may seem like one of the last countries on earth to offer lessons for dealing with economic malaise. Once the eighth-largest economy in the world, it steadily slid through the 20th century, thanks to decades of repressive dictatorships and inconsistent market experiments. This ended ignominiously in 2001, when it defaulted on $100 billion in sovereign debt, plunging over half its 35 million people into poverty. . . . Since then, it has performed an economic U-turn. . . . Argentina is not without problems, but its recent economic record speaks for itself. The economy has grown by over 6 % a year for seven of the last eight years, and unemployment has been cut to below 8 % today from over 20 % in 2002. The poverty level has fallen by almost half over the last decade. The streets of Buenos Aires are choked with cars as Argentines are on track to buy some 800,000 new vehicles this year. The wine mecca of Mendoza is full of high-end tasting rooms, hotels and restaurants offering regional haute cuisine. Plasma TVs and BlackBerrys have become household staples among the urban middle class. Argentina has regained its prosperity partly out of dumb luck. A commodity price boom has vastly benefitted this soy, corn and wheat producer." (Mount, 2011)

Mount also wrote: "But it has also prospered thanks to smart economic measures. The government intervened to keep the value of its currency low, which boosts local industry by making Argentina's exports cheaper abroad while keeping foreign imports expensive. It then taxed those imports and exports, using the money to pay for a New Deal-like public works binge, increasing government spending to 25 percent of G.D.P. today from 14 percent in 2003. As a result, the country has 400,000 new low-income housing units, as well as a long-delayed, 235-mile highway between the northern cities of Rosario and Córdoba. It has also strengthened its social safety net: the Universal Child Allowance, started in 2009 with support from both the ruling party and the opposition, gives 1.9 million low-income families a monthly stipend of about $42 per child, which helps increase consumption. Because the amount depends in part on how often the child attends school, it is also likely to improve the country's long-term educational performance. The results have also paid off politically: President Cristina Fernández de Kirchner recently won about 50 percent of the vote in an open primary against nine other presidential candidates. . . . Of course, Argentina is far from perfect: the import and export taxes have scared away some foreign investment, while high spending has pushed inflation well over 20 percent. There are also problems with the way Argentina is run: corruption, government opacity, authoritarian tendencies, confiscatory taxes and a temptation to tweak unpleasant inflation statistics. . ." (Mount, 2011)

Lessons Learned

Modern nations depend upon foreign investment to fund industrial developments in the nation and to fund government infrastructure developments. When governments use foreign investment to principally fund government services, the tax base of the country has been insufficient to balance the government budget. Deficits continue to grow, year after year. When the annual government budget deficit becomes a significant portion of the annual revenue of an economy (GNP), then foreign investors lose confidence in a government's capability of continuing to finance its deficits. Investors stop buying government securities. The national currency exchange rate plunges. Local banks in a nation become insolvent. Bank runs occur as savers withdraw deposits from banks. Credit stops in a national economy, and business are unable to finance day-to-day production and pay wages. The economy plunges into a depression. Masses of people are unemployed. Property is lost. Families starve. Governments fall. The society descends into chaos.

International financial institutions had neither correct economic models nor effective policies nor proper regulatory power to ensure that the global financial system was sound and also beneficial to economic growth in the nations of the world.

The macro-economic theory for international development was merely an export-import ratio without any national context. Why macro-economic theory had drifted toward a lack of economic context has been commented upon by Peter Boettke. He wrote: "F. A. Hayek's realistic economic theory had been replaced by the formalistic use of equilibrium models, that bear little resemblance to reality. . . . The first casualty of the formalist revolution (in economics) was the historically and institutionally rich tradition of economics still evident in the 1930s. Case studies of particular industries, for example, had been common. After the development of econometrics, however, the case-study approach was discarded in favor of large-sample data analysis. The second casualty of the formalist revolution was what might be called "the economist's way of thinking," the defining characteristic of the discipline in both its classical and early neoclassical renditions. (This formalistic thinking lost) the best of the earlier economics, which had combined an appreciation for the particularities of institutional context – with theory grounded in the generalities of choice under conditions of scarcity." (Boettke 1997).

In reviewing both the IMF Polak model and the actual cases of Argentina and Korean experiences with IMF policy and their own recoveries from the 1998-99 recessions, we saw evidence for Boettke's criticism of a 'formalized macro-economic theory without context'. The IMF policies should have instead focused not just on 'macro-economics' but also on the 'micro-economics' of each nation – *a macro-economics with a micro-economics context.*

The IMF macro-economic policies focused only on increasing credit interest rates, closing banks, increasing taxes, reducing government expenditures, and increasing unemployment. In this way, IMF officials believed that exports would be automatically increased and imports automatically decreased and international monetary exchange rates would be automatically adjusted. What was the basis for this belief? Joseph Stiglitz wrote: "There is little doubt that the IMF and Treasury policies contributed to an environment that enhanced the likelihood of a crisis by encouraging, in some cases insisting, on, an unwarrantedly rapid pace toward financial and capital market liberalization." (Stiglitz, 2002)

Markets need to be properly regulated to approach market perfection. Ideally 'markets' should be controlled by prices, set when supply meets demand in a market. But this happens in real markets, only under properly regulated conditions, such as: no monopolies, safe products, honored contracts, no business fraud, no government corruption, etc. What the IMF economists should have been paying attention to were the 'restrictive conditions' that foster ideal markets – the 'priors' of effective economy, the microeconomic structure of a nation.

The Stiglitz conditions for markets to operate economically ideally in a country are: adequate regulatory framework, adequate competition, safety nets for economic liberalization, and policies for job creation rather than destruction. To these, we can add that governments should collect sufficient taxes to fund the needed government services for a healthy societal infrastructure.

Governments should borrow money from foreign financial markets principally for finite infrastructure building projects and not for funding current government services (with the exception of in a depression, a government may need to stimulate restarting a depressed economy, a la Keynes).

In comparing the two crisis experiences of Argentina and South Korea, one can see that a micro-economic structural context (details of the nation) is necessary and empirically essential – for the proper use of macro-economic theory, in fixing a national financial crisis.

Macro-economic theory, by itself, cannot explain national economic growth; micro-economic theory (as the detailed structure of industry and its proper relation to government) is necessary both to explain and guide policies for economic growth.

Chapter 9

HIGH TECH SME – NANO

INTRODUCTION

As we have seen for the large companies in Korea (and in the world), technology is an essential competitive factor. Any business must have technology (in products, production, services, or operations) equal to any other competitor, or it will suffer a competitive disadvantage in technical performance or quality or cost of products/services. This is defensive, having technology equal to that of competitors. Offensive strategy is to have technology superior to competitors. We saw in the case of Samsung Electronics, it jumped ahead of competition to gain superior technology, in next-generations of memory chips); and global competition was beaten by Samsung. New technology is a necessary competitive advantage for large companies in the global economic world. So too is it necessary for small to medium sized businesses, SMEs. But large companies have expensive R&D laboratories to assist them in gaining new technologies. Where can SMEs find and gain new technology, as a high-tech SME?

A high-tech SME is one innovating new products and/or production with new technology. There are several types of SME businesses: retail, service, and manufacturing. We focus here on the high-tech manufacturing type of SME. For it, the challenge lies in new technology for new products and for new production. As SMEs often have limited R&D capability (due to their smaller size), from whence can new technology come? An important source is university research.

HISTORY – NANO CORPORATION

We recall that 1999 was a difficult year in South Korea because of the Asian Financial Crisis. The new President Kim Dae-jung, issued a policy to encourage university professors to start companies to provide employment for graduates. The government allowed professors to take patents on their research. The government also allowed universities to provide space for professors to begin businesses. And the government also provided a grant program to fund technology development for new small businesses.

At the time, Shin Dong-woo was a professor of materials science at Gyeongsang National University. He had been doing research in nano-particle processing and had invented a new process for producing very small titanium-oxide (TiO_2) nano-particles (20 nanometers) at a cheap cost. These were ten times smaller than commercially-available titanium-oxide particles (then at the size of 200 nanometers). The nano-particles of TiO_2 would be much more efficient in the application of these particles to act as a chemical catalyst because the smaller particles could provide, all together, a larger surface to assist (catalyze) chemical reactions.

To make electrical power, many power plants burn coal or oil. The burning produces an exhaust gas with a poisonous component of nitrogen oxide. This needs to be removed from the exhaust. It is removed by first passing the exhaust fumes through a catalytic filter, before going out into the environment. The chemical reaction in the filter converts the nitrogen oxide fumes into nitrogen gas and oxygen gas. To accomplish this helpful chemical reaction, a catalyst is used. A catalyst in chemistry assists a particular chemical reaction to occur faster and at a desired temperature. The catalyst in the filter is a molecule made up of titanium (Ti) and oxygen (O), called T_iO2 (titanium dioxide). The presence of this catalyst allows the reaction of nitrous oxide (NO_x) with ammonia (NH_4OH) and converts NO_x to nitrogen and water, at the low temperature of the flue gas. Otherwise, the flue gas would have to be reheated to a high enough temperature for the reaction to occur without the TiO_2 catalyst. Shin applied for and received patents on his process in Korea, in the United States, and in China – to inexpensively make TiO_2 nano particles.

Professor Shin then started a new business to produce nano-particle TiO_2 as a product and also to employ his graduate students, majoring in Materials Engineering (who could not in 1999 find other employment in Korea‘s faltering economy). His called his new company, Nano, and applied to the Ministry of Environment for a grant of $200,000 to build a pilot plant, in a building of 200 square meters, owned by the university.

NATIONAL INNOVATION SYSTEMS

A key idea underlying high-tech SME's is the concept of a national research system – in which industrial technological research capability is connected to university research performance. In this case of a Korean new venture SME, we will see the importance of several kinds of policies for connecting university research to new high-tech ventures – policies in government, in universities, in banking, and in faculty members as entrepreneurs.

In 1994, the term 'national innovation system' was used by Chris Freeman and by the OECD to describe the national infrastructure which facilitates research and development (R&D) and its contribution to technological innovation. (Freeman, 1994) (OECD, 1997) In 1998, one of the authors depicted the societal infrastructure of such a system in Venn diagrams, Figure 9.1. (Betz, 1998)

Research performed in the infrastructure of a national innovation system is transformed into utility as new products/services. This infrastructure operates at two levels, a macro-level of innovation and a micro-level of innovation. The macro consists of three institutional R&D sectors – university, industry, and government – each providing research activities toward inventions. Below that, a micro-level consists of the set of high-tech businesses which produce new products/services that embody newly invented technologies. The national innovation process consists of the procedures by means of which the macro-and micro-level organizations interact – to create radical technological innovation. The design of high-tech products/services using the new technology is the way that new knowledge is transformed into new economic value – as the goal of the knowledge-to-value transformation.

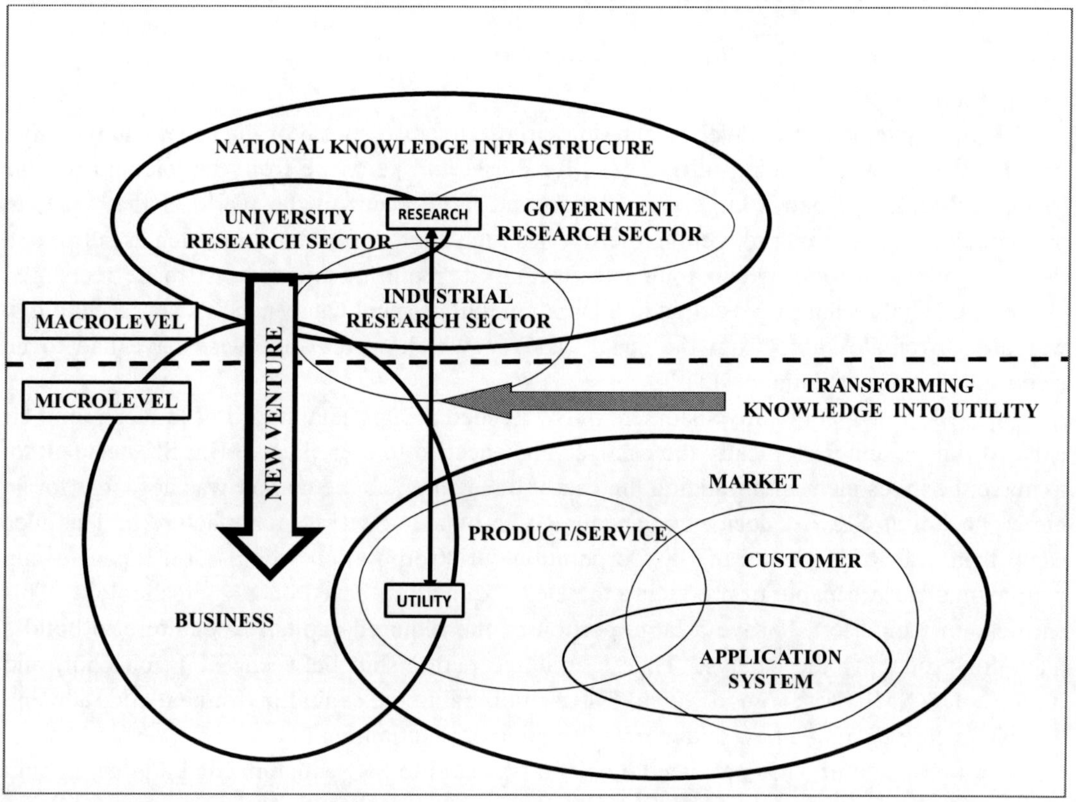

Figure 9.1. National Innovation System.

History – Nano Corporation (continued)

In 2000, Nano successfully completed the pilot plant, producing a powder of TiO_2 nano-particles. Nano acquired $5,000,000 in venture capital from investors to build a full-scale production plant.

Professor Shin first contacted a fund manager who worked for the Korea Development Bank (KDB), which was then the largest state-owned bank in Korea. In 1999-2000, an IMF rescue fund was supplied to Korean government, which was transferred to governmental banks, such as KDB. The manager of the fund in KDB was an invited foreign manager from Britain (who had prior accomplishments in this field). This British fund manager was looking for companies in which to invest. But demand for the venture capital was diminished due to the economic troubles, and there were few healthy start-up companies, resulting in a temporary surplus of venture capital. Also it was helpful that the CEO of Nano was a Ph.D. graduate of Cambridge University. The initial investment of KDB fund to Nano was $2 million. Then KDB became a share holder of Nano, helping as a strong and stable financial supporter. This, along with the high reputation of British fund manager, helped Nano raise an additional $3 million from other Korean venture capital, to build a full-scale plant.

In 2010, Nano received its first order for ten tons of titanium-oxide nano-particle powder from SK. SK Group is one of the Korean conglomerates and the leading the energy and chemicals firm. At that time, SK was looking for a partner to manufacture this catalyst for them, a Selective Catalytic Reduction (SCR) DeNOx catalyst, that would use the TiO_2 nanoparticles.

SK had developed a recovery technology to use for its disposal of the spent catalyst from its own process of producing ultra-low-sulfur diesel and kerosene from the vacuum residue (VR) feedstock. SK had a huge unit (the second largest unit in the world at that time) to produce this. And in that production, there was a high content of impurities such as sulfur and metals in the feedstock, which fouled their catalyst, requiring the catalyst to be recharged every year. That recharging resulted in a large amount of spent catalyst. SK's new technology was able to create value from the spent catalyst, which would otherwise have had to be disposed of at a substantial cost.

In 1998, power-company customers of SK needed a TiO_2 catalytic filter to clean up their exhaust fumes, and periodically the catalyst TiO_2 needed to be replaced. But SK decided not to make the investment in producing the catalyst and filter, as the market was not large for so big a chaebol as SK. SK decided to encourage an outside SME to manufacture it. This idea came from the R&D Center of SK Corporation. SK Corporation would form a partnership with a small firm capable of producing the DeNOx catalyst, based on SK's technology. This partnership would let SK save a large portion of the required capital expenditure to build a plant for producing the catalyst. Thus a strategic partnership between SK (a chaebol) and Nano (a new SME) was a good way to SK of minimizing the capital investment and reducing the risk in constructing a new commercially-viable production facility.

The first batch of TiO_2 produced by Nano was sold to SK, which needed a large amount of TiO_2 in test-producing the SCR DeNOx catalyst product, based on their own technology. Successful test production led to the signing of the partnership agreement between the two companies. At the same time and in the same company, the future Nano had found both a partner (SK) and a customer (SK) for their nano particle TiO_2 product. And Nano had invented the technology for the new product (nano-particles of TiO_2) in prior university research.

SK Group was founded by Chey John Hyun, when in 1953, he acquired Sunkyung Textiles, which later in 1958 produced Korea's first polyester fiber. In 1973, SK began Sunkyong Oil in 1973 to produce its own source of materials to produce the fiber. In 1980, SK added to it petroleum refining capacity by purchasing Korea National Oil. In 1994, SK entered the telecommunications business.

INNOVATION THEORY: NEW VENTURES

Nano was one of the many new companies begun from new technologies, a characteristic of the innovation processes in the twentieth century. Scholars studying the innovation phenomena of new start-up high tech companies had come up with a theoretical pattern, which has been called 'new venture theory'. In a successful new venture, there are five key ingredients: a new product/process/service idea based on new technology; the organization of a new firm; an initial venture team; initial capital; and focus upon a potential market. (Betz, 1987) This theory does fit the case of Nano: Professor's Shin's invention of nano-particle-sized TiO_2 powder was a new product usable as a catalyst substrate; he founded Nano as a new business; he and his graduate students made up the initial team; he located initial capital as a grant to build a pilot plant; and he focused upon a potential market (in the industrial process of cleansing exhaust gases from coal-burning electrical power plants).

Nano fits traditional new-venture theory, but with some important differences. First, it is unusual that a professor and his graduate students formed the initial team. In most cases of the U.S., venture teams are composed of former employees of big companies. In cases of new ventures from university research, often it has been the graduate students who formed their own company. In this case, the professor was encouraged by the government, to start a new venture with graduate students. This has not often been a formal policy of governments. Second, another difference is that the government allowed university professors to own patents on their inventions created during university research. Usually the university owns patents on professor's inventions, and a professor may or may not be allowed a license. Third, a difference is that the initial capital came from a government pilot-plant grant (and not from venture capital). Fourth, the university provided initial space for constructing the pilot plant. Since the 1980s, many universities have provided incubator spaces on or near their campuses.

Thus in this case, some aspects of traditional new venture theory must be modified, in order to facilitate new ventures by professors directly from their university research – professors as venture leader with graduate students as part of the venture team, intellectual property owned by the professor, government research grants for technology development, and university-provided incubator space.

History – Nano Corporation (continued)

There were two products for Nano to make. The production of TiO_2 powder at the nano size was one product. The production of a plastic honey-comb-shaped filter to hold the powder was a second product. Nano first built a facility to produce the powder and then built production equipment to construct the honeycomb frame (called SCR), which held the catalytic titanium-oxide nano-particles. In 2003, supplying both catalytic powder and honeycomb SCR filter, Nano fulfilled its order to SK. However, the year before in 2002, Nano also had supplied its honeycomb catalyst to the Bundang Complex Power Plant (before Nano and SK signed a partnership agreement in 2003). So Nano began with not just a single customer of SK but also a power plant customer. This was a first step toward Nano avoiding having only one large company as a single customer.

Bundang is a public company which supplies heat and power to the Bundang municipality located in the south of Seoul. It is a small power plant and does not require a large amount of DeNOx catalyst. Nano was able to supply this company with its own product produced by the pilot facilities. This was before the partnership agreement was signed between Nano and SK Corporation. This indicates that Nano had its own "recipe" to manufacture DeNOx catalyst without using the powder and the recipe using the spent catalyst from the heavy oil desulfurization unit. Thus there was no problem about intellectual property between SK and Nano.

Nano thought, at the time just after the Bundang project, that it needed more than just catalyst-making technology to become a serious contender in the DeNOx catalyst market. It needed the appropriate production capacity for quantity, quality, and cost. If a vendor wanted to supply DeNOx catalyst to KEPCO (Korea Electric Power Company), the single largest customer in Korea, it had to meet strict eligibility requirements: a capability of supplying at least a 200 Mega Watts DeNOx replacement capacity and also having had one-year-or-longer continuous operation. Only foreign companies with long history of manufacturing and marketing were allowed to bid for the DeNOx catalyst, because of the stringent requirements. Nano was determined to meet them and become a DeNOx catalyst supplier.

In 2003, Nano also supplied its first major export of the honeycomb SCR catalyst to EnBW in Germany. This effort required overcoming huge production problems in the new product. It was delivered on time (by flying the very heavy product to Germany). There the TiO_2 nano-particle powder was applied as a catalyst substrate in the stream of exhaust gases from coal combustion by having a honeycomb material, providing an SCR catalyst.

Thus the president of Nano, Professor Shin, had decided not only to produce his patented nano-particle powder but also to make the catalytic honeycomb to insert into the exhaust gas stream. To make the honeycomb, Nano had to develop its own extruder equipment with which to extrude the honeycomb from dough materials. Nano had developed two kinds of production processes: to produce the nano-particle powder and to extrude the honeycomb.

During 2004, Nano continued to improve both production processes, producing powder and extruding the honeycomb filter. These development efforts were helped by collaborations between NANO and a couple of other Korean SME's. Another research grant of $2.5 million from the Korean government enabled NANO to fund research for continuing to improve its production technology. Figure 9.2 shows Nano products.

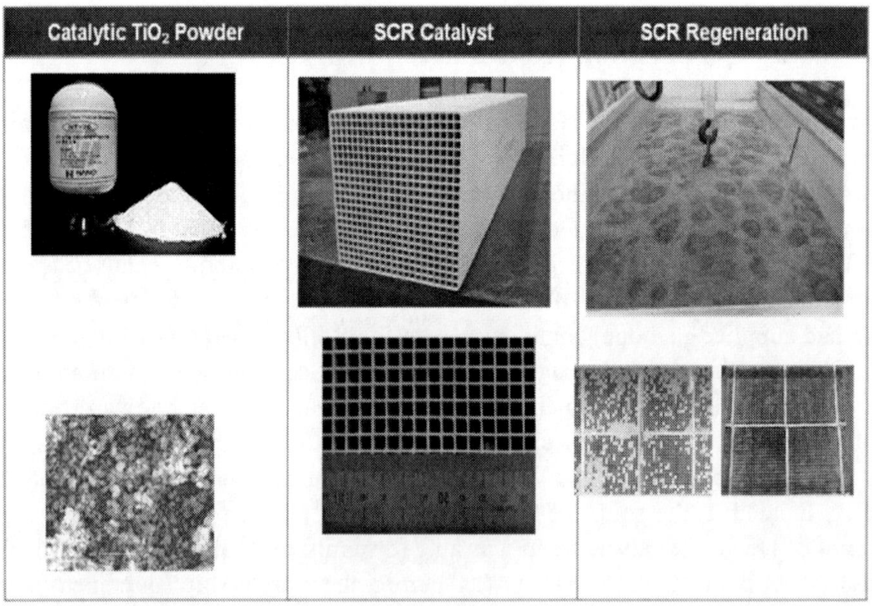

Figure 9.2. Tio_2 Catalyst Powder and Scr Catalytic Filter.

INNOVATION THEORY: NEW VENTURE MILESTONES

Another part of the theory of new ventures consists of the milestones which new ventures have to meet: Acquisition of start-up capital, Development of new product or service, Establishment of production capabilities, Sales growth, Production expansion, Meeting competitive challenges, Product improvement, Production improvement, and Product diversification, Organizational and management development, Capital liquidity. Nano would meet all these milestone challenges.

Acquisition of Nano's start-up capital was, as we recall, the initial KDB fund decided by British fund manager, and the funding by highly reputed British fund manager readily attracted other venture capital up to $5 million, within a month. With that money, facilities were built for the production line for TiO_2 (Nano's main product at the initial stage) and to manufacture an SCR DeNOx catalyst. This was a fully integrated unit, although of pilot plant scale. The first batch of DeNOx catalyst supplied to Bundang Cogeneration Plant was produced using this pilot-scale unit. The building for the pilot plant facility had been provided by the university.

For the development of a new product, the TiO_2 and DeNOx catalyst of Nano's own manufacturing recipe was developed using the facilities and by employees, most of whom were Professor Shin's graduate students. For the development of processing machines and testing and evaluation equipment, NANO took advantage of governmental R&D fund which lasted for four years from 2002. The total R&D fund supported by government was $2.5 million. During this initial period, Nano followed two development paths. One was the production of catalyst for captive user, SK (using the hardware provided by SK and Nano). Second was the development of improved cost-effective technologies by Nano, for processing, monitoring, quality-control, and production machinery.

For the establishment of production capabilities, we recall that Nano had the capability to produce DeNOx catalyst on its own, even before the SK-Nano partnership was formed. Nano simply had to produce DeNOx catalyst based on the SK recipe, which was a key element of the collaboration agreement.

For sales growth, even with the collaborative effort with SK Corporation, initially the rate of sales was slow and disappointing. This was because of the stringent requirement by the KEPCO. The breakthrough for Nano occurred when ENBW, the third largest power company in European Union, decided to buy the SK-Nano product. The decision was made after a series of bench-scale tests at ENBW and all-out marketing effort by SK Corporation. ENBW, after being fully convinced by the SK's impressive R&D capabilities and with a heavy price discount offer from the Korean partnership, finally decided to buy roughly one-million dollars worth of product. SK-Nano partnership returned ENBW's favor by delivering quality products on time. Toward the last stage of the product delivery, SK Corporation paid a high air freight bill to deliver the DeNOx catalyst on time for the power plant start-up. With the reference to ENBW sales and satisfactory performance records at the German power plant, SK and Nano became the sole eligible domestic company to sell DeNOx catalyst to KEPCO – the single, largest Korean user of the product. Soon, the market share of the SK-Nano DeNOx catalyst soared to over 60% in the domestic market. Here SK (a large conglomerate) helped Nano (a new SME) to find a significant first customer, other than SK.

For production expansion, Nano needed twice to build a new plant. Nano's choice for its first DeNOx manufacturing plant was to locate it in the suburban area of the southernmost city called Jinju (which is still known as the city of school teachers because of the high ratio of teachers-to-population). All the professional employees and the CEO of Nano were either graduates or professor at the National University of that small city of education. But with continuing sales growth, both in the domestic and the foreign markets, there arose a need to expand Nano's original 2,000 cubic meters/year capacity to a higher level; and this became a crucial issue in the partnership with SK. But the interest in the partnership by SK waned. The manger of SK Corporation, who first encouraged SK's participation, had moved to another position in the SK Group CTO. His successor not interested in strategic alliances with SME companies. One issue at the time, was whether SK might be interested in helping to finance production expansion in Nano. SK then was not interested. That partnership, which had lasted for over 8 years, ended. NANO was confident on its independence, since Nano had developed by itself all subsequent production technologies and equipment. Nano needed to expand its capacity, looking in areas outside Jinju, as potential sites for a new plant, quadruple the size of the first plant in Jinju. The city of Sangju, located in another province, made an extremely favorable offer to Nano, expecting that Nano would bring jobs and tax revenues to that city. Two banks, IBK and KDB (which are essentially owned by the government), were competing against each other to become the fund provider to Nano. This is partly due to the policy of the Korean government, who wielded a great influence on those two banks. President Lee of the Republic of Korea was at that time pushing for strong SME policy in order to overcome world economic crisis in 2008-2009. Everything in government policy about SMEs was continuing to be in Nano's favor at that time.

Meeting competitive challenges required that Nano face competitive challenges in both domestic and foreign competitors. There were no serious domestic competitors for Nano for domestic markets. Since the financial crisis which had hit Korea in 1998, several Korean companies attempted to establish themselves as viable players in the domestic market. None of them were able to match Nano's capabilities in terms of product quality, competitiveness in cost and production capacity.

A couple of much smaller companies are competing against Nano, but only in the very small market of cogeneration and garbage incineration plants. In the overseas market, the situation was different in that a number of companies with sizable production capacity and cost competitiveness were fighting for the market share. There was a US-Japanese joint venture called Cormetec, and most of the serious players in the DeNOx catalyst market were European companies. But by 2013, Nano was almost the sole player in the DeNOx catalyst market that had expanded the production capacity (and which then appeared in 2013 to have a lot to do with the depressed economies of both European and Japan). Also in 2013, the sector with the prospect of large growth in the DeNOx catalyst market was next in the marine application area. Nano has been aggressively becoming the leader in this emerging area, whereas it had been a follower in the electrical power market. The fact that Korea is one of the leading countries in the global shipbuilding industry may put Nano in a favorable position for the new DeNOx market for marine application. One German company, which designs the SCR DeNOx system for the big ships propelled by large-size diesel engines, has formed a partnership with Nano in that emerging marine DeNOx market. This indicated the excellent reputation Nano had achieved in the international market.

For product improvement and production improvement, Nano has steadily improved its product quality to such a point that it can satisfy the quality expectation by the sophisticated European electric power generators as well as those of Asian customers including Japanese companies. One of Nano's accomplishment was in production improvement. All the equipment and the facilities associated with the DeNOx catalyst manufacturing are manufactured and supplied by the Korean SME companies that have close working relationships with Nano. R&D funding from the Korean government played a critical role in letting Nano and the domestic partner SME companies come up with a complete set of technological capabilities to fabricate essentially all of those used in the process of DeNOx catalyst manufacturing. This has put Nano in a position to compete against any foreign catalyst manufacturing companies.

For the product diversification, Nano developed a regeneration technology which allowed the waste catalyst to be used again. This technology has been competed with fresh catalyst module, but accepted by most of Korean power plants due to half cost. It was a lucrative business in 2010-11 before many of market followers duplicated a similar technology. In addition, Nano steadily developed the catalyst for marine application, which required high resistance to vibration, and became global top supplier for marine market since 2012. Nano has also developed the plate type SCR catalyst which occupied 50% of domestic and world SCR market, with only three other companies in Japan and Europe competing to manufacture it. A new production line for the plate type SCR catalyst with 10,000 cubic meter capacity will be completed by the end of 2013. At the same time, the production line for honeycomb catalyst for marine application will also be increased up to 6,000 cubic meters by June 2013.

In 2013, Nano was still classified as a SME. Nano would need to increase its sales revenue and the profit before it attempts to hire qualified R&D work force for further product diversification. But for the existing business arena of power plants and the marine application, Nano's work force is more than qualified to compete against the bigger companies in Europe and Japan.

INNOVATION THEORY: INDUSTRIAL VALUE-CHAIN

We next review more theory about the management of technology which applies to the case of Nano, which is the concept of an industrial value chain. (Porter, 1985) and (Steele, 1987) The traditional theory of the industrial value chain indicated suppliers (vendors) to companies further along on the value chain; but it had not specifically mentioned strategic business alliances between suppliers and industrial customers. Yet in this case, one sees the importance in the beginning of a strategic alliance between SK, a large conglomerate firm (Korean chaebol), and Nano, a new SME high-tech supplier. The alliance was to produce a product for an SK. SK did not wish to invest in producing that product, as the market was seen as not large enough for SK. But for Nano, this market would be very large. Without opposition from SK, Nano continued to build their business with ex-SK customers (e.g. EnBW). Thus to the traditional theory of the industrial value chain, one can add the idea of possible 'strategic business alliances' between a large conglomerate and a new high-tech SME, as shown in Figure 9.3.

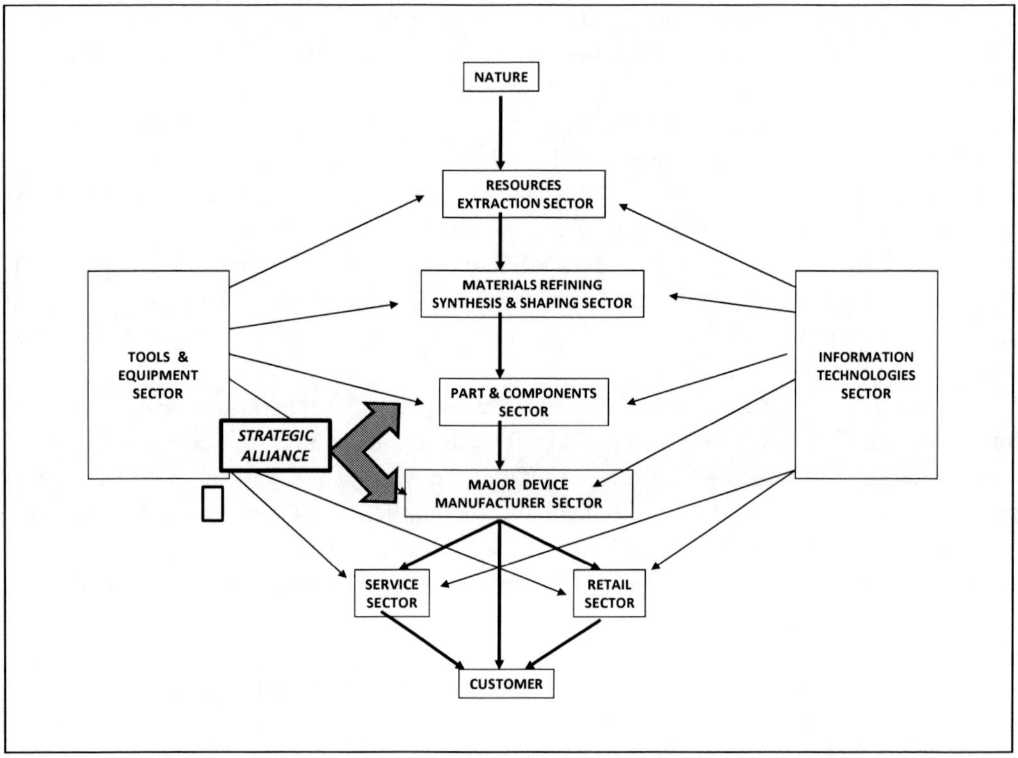

Figure 9.3. Industrial Value Chain – Strategic Business Alliances between Conglomerates and Smes.

History – NANO Corporation (continued)

As a new business, Nano grew by adding new customers. In 2005, SK-Nano supplied a honeycomb SCR to a Korean Coal Fired Power Plant in Boryong. In 2007, Nano supplied a honeycomb SCR to a Heavy Oil Power Plant in Ulsan, independent from SK. Also it supplied honeycomb SCRs to Diesel Power Plants in Jeju Island.

In addition to new customers, Nano expanded its business by adding services along with products, In 2008, Nano began providing catalyst maintenance services for replacing catalysts in SCR honeycombs, with a first customer at Samchunpo Power Plant.

To help finance this service expansion, new research and development (R&D) efforts had been made and initiated in 2002, partly with the government money. It developed the equipment for all the production (equipment, extruders, dryers and other critical units), in order to reduce the production cost and control quality.

As business expanded, Nano needed to expand production capacity. In 2009, Nano built an additional facility for producing SCR catalyst at Sangju at the area of 50.000 square meters using production equipment developed by Nano.

Nano had begun with research from the university and then continued research to improve production capabilities. Nano remained high-tech in R&D. In 2009, Nano hosted a 3rd International Conference for SCR maintenance technology, broadening its exposure and contacts with the technology community. Again using government research funding to help finance its research, Nano received another research grant in 2009, from the Korean government, to develop a SCR catalyst for marine applications.

In 2010, Nano's success at business and high tech was recognized, as Nano received a presidential award for top technology. It was selected by the Korea Credit Guarantee Fund as one of Korea's top five green technology companies. Nano hosted a 4th International Conference for SCR Technology in Busan. Nano was appointed as candidate of hidden champion enterprises by Korea Export Import bank.

In 2012, Nano built a new R&D center at Sangju, continuing R&D and generating new SCR patents. Nano completed a second production plant in nearby Sanju. Nano hosted the 5th International Conference for SCR Technology in Busan in June, and Nano opened an office in Shanghai, China: NANO CHINA CO.

Also in 2012, Nano anticipated a growing market in applying SCR technology to diesel powered commercial ships. Nano aimed to become a leading manufacturer in the emerging SCR market for marine applications. Korea was one of the dominant ship builders in the global market, which is help Nano's position in this newly opened market. The ship market could become larger for Nano than the existing SCR market for the coal-fired power plants.

In 2013, a third large investment in Nano history is scheduled to increase the capacity of honeycomb production up to 6,000 cubic meters by June, and a new establishment of plate type catalyst with annual capacity up to 10,000 cubic meters by the end of 2013. A 6th International Conference for SCR Technology will be held in Busan in September in 2013. Nano's revenue in 2013 is expected to be $50 million, with a projected $100 million in 2014 (due to the a new big market formed in China from 2013). In 2013, 90% of Nano's revenue was expected to be from export to foreign markets.

INNOVATION THEORY: INNOVATION HELIXES

In 2000, H. Letzkowitz and L. Leydesdorff introduced a time-dimension of interactions within this national system infrastructure, calling such interactions as a kind of "triple-helix of university-industry-government relations". (Etszkowitz and Leydesdorff, 2000) In 2009, E.G. Carayannis and D. F. J. Campbell extended this triple-helix model to a quadruple-helix by adding another interaction in the innovation process of the civic public, as media and culture. (Carayannis and Campbell, 2009) (Carayannis and Campbell, 2010) In 2012, they expanded upon the elements of the fourth helix of civil society and culture, as shown in Figure 9.4. (Carayannis and Campbell, 2012)

This depiction of the four helical forms in the processes of innovation emphasize that these institutional processes occur in the three research sectors of government, university, facilitated by a civil society and culture. Also it emphasizes that they are all cyclic in two ways: cyclic in knowledge processes in each sector and cyclic in passing knowledge between sectors, each respectively focused upon 'funding', 'science', 'technology', and 'culture'. In this case study, we have examined the interconnecting processes for innovation by a new SME in Korea. Nano Corporation was a new venture, beginning in South Korea in 1999 directly from university research.

Also a particularly interesting quadruple helix innovation process which involves SMEs is one prevalent in Germany in what has been called 'Mittlestand'. Mittlestand is the German term given to the German small-medium-enterprise (SME) businesses which have been noted for their contemporary strong competitiveness in the global market – due to corporate strengths (family businesses) and technological strengths (technical quality in niche markets). There have been many studies trying to understand the reasons for the German success, such as: (Meyer-Stamer and Wältring, 2000), (Günterberg and Kayser, 2004), (Venohr and Meyer, 2007), (Venohr, 2010). These referenced an earlier study (Simon,1996). The interest in the German Mittlestand has increased as they have significantly contributed to Germany's strong export economy. And as Carayannis and Campbell emphasized, such a Mittlestand culture fits into the fourth strand of the quadruple helical innovation processes, national civic culture. In this case study, we also examine how it could be used as an exemplar for a growth of a Korean kind of Mittlestand. Nano emerged as an SME, which was a major exporter of new high-tech products in the global market, as a supplier to energy production businesses.

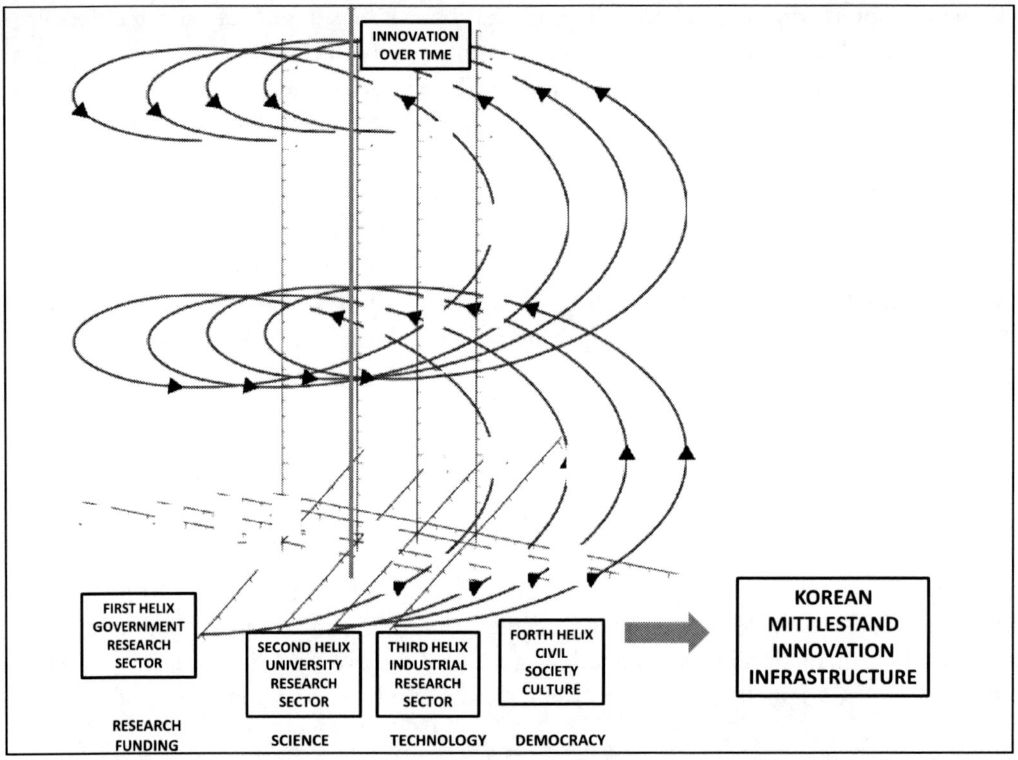

Figure 9.4. Interacting Cyclic Innovation Processes: Quadruple Innovation Helixes.

At the time of its founding, Nano was not technically a 'family business'. But it was founded with a kind of 'intellectual family' in mind. Professor Shin specifically began the company to provide employment for his materials engineering graduate students, and his former graduate students still form the upper management-level of the new company. Nano focused toward global market dominance in the SCR market for coal-fired electrical power plants and next for diesel-powered commercial ships. Nano had continuing R&D efforts to improve operational effectiveness (in its processes of producing Nano-particle catalytic powders and in extruding honey-comb filters). However, the Korean education system has not had a German focus upon vocational training and the equivalent of the German Frauenhofer

Institutes have not yet been institutionalized in the Korean national innovation system. Only three of the five Mittlestand features can be found in the case of Nano's early success. (And we will discuss this in the next chapter.)

However, Nano has also kept a unique educational system, which could compensate for the gap between Korea and Germany in the respect of vocational training. Nano consistently encouraged employees to attend the master and Ph. D courses at the university, wherein Professor Shin continued as a faculty member. Their theses for advanced degrees were relevant to the product of NANO. Their experiments were carried out at Nano R&D centre. The time and tuition fee for advance degrees were fully supported by Nano as company policy. As a result of this part-time degree program, three doctorates were graduated. They began working in Nano as heads of production, R&D and raw powder materials divisions. And there were ten other employees with masters degrees in various majors, including management. This part-time degree program was still operating with one Ph.D. candidate and two masters in diversified fields.

Lessons Learned

This case offers evidence that the four strands of helical innovation interactions are all important. Firstly, that university and industrial and governmental research sectors interact in innovating new high-tech SME businesses. Secondly, government research funding agencies should have programs to sponsor research grants from SMEs to improve technical processes. Thirdly, strategic technology alliances between SME suppliers and large conglomerates are important to grow a new technology market, which may be too small for the conglomerate but very significant for the SME. Fourthly, the proper civil culture and proper government policies are important to encourage university professors to establish new high-tech businesses from their university-based research.

As seen in the case, most aspects of new venture theory still apply across America, U.S. and Asia. And in the theory of national innovation systems, all high-tech enterprises require national research and development (R&D) infrastructures and processes, in order to begin and to continue to be technically competitive in quality and cost. Large corporations can afford the expense and difficulty of running research laboratories and also of connecting to university research. However this is especially difficult for SMEs. In the case of Nano, university research provided the new next-generation technology upon which the business was built. Also government policy had encouraged university professors to spin new businesses out of their university research. Government policy allowed the new business the patent rights on the professor's invention, upon which the new venture was launched. Also there was a government research funding program accessible to Nano, to finance further research to improve Nano's technical processes.

In the theory of the German Mittelstand SMEs, Bendt Vehnohr emphasized the strategic positioning of the German Mittlestand: "Strategy – Dominate Global Niche Markets; Governance – Enlightened family capitalism; and Operational Effectiveness – World class in key processes". (Vehohr, 2010) In the case of Nano, similar strategic positioning occurred, as Nano was focused upon the global niche market of TiO_2 catalytic filters for combustion emissions; Nano formed its governance with a focus upon employment for Korean materials engineering graduates; and Nano continued R&D to improve operational effectiveness in products/processes. The next issue is whether South Korea could adapt the model of the German Mittlestand SMEs to Korean culture?

Chapter 10

EMPLOYMENT AND MITTLESTANDT KOREA

INTRODUCTION

Compared to the large companies, the Small-Medium Enterprises (SMEs) in Korea have encountered their own economic problem, also common to SMEs in most other countries. Global companies operating in a nation, such as Korea's chaebols, provide the majority of Gross Domestic Product (GDP) revenue but only a minority of jobs. In contrast, a nation's Small-Medium Enterprises (SMEs) generate a minority of the GDP revenue but provide the majority of jobs. In 2012 in South Korea, chaebols generated 75% of Korea's GDP and provided only 25% of employment. Conversely South Korean SMEs generated only 25% of GDP but provided 75% of employment.

For most nations (except Germany) in the world, this was the general pattern: global enterprises created the wealth, while SMEs provided employment. It was one of the serious paradoxes in global civilization: wealth versus employment, big versus small businesses. For example in 2011-2013, there was a Euro-crisis in Europe. In most European nations, global corporations paid little taxes and unemployment rose to depression levels. The exception was Germany, which did not suffer a Euro-crisis, had a positive trade balance and low unemployment. The distinguishing economic feature of this German successful economy was that, in 2010, German SMEs generated 50% of GDP and 60% of the employment.

Except in Germany, wealth versus employment is a serious problem in the modern global civilization. As we studied earlier in the case of Nano, we will address this SME problem as it existed for South Korea in 2013 – again using the German model for SMEs, the 'Mittlestand'.

HISTORY – PRESIDENT PARK GEUN-HYE

As we earlier reviewed, the 1997 Asian Financial Crisis had a major impact upon South Korea, with the government closing down some insolvent chaebols, reforming the financial position of the remaining chaebols, and embracing the start-up of new high-tech SMEs, such as Nano. And as we saw in an earlier chapter, the Samsung group restructured its finances (selling off Samsung motors) and concentrated upon its electronics businesses. By 2010, Samsung gained 50% of the global semiconductor memory chip markets. Also Posco began to develop steel production capacities abroad, building plants in India and Indonesia. In 2010,

Posco was the fifth largest global steel producer. Hyundai acquired Kia and continued to build its auto industry. In 2010, Hyundai was the fifth largest automobile company.

Thus all was well with the Korean economy, correct? Not quite. All was well with many of the Korean chaebols, but not completely with all Koreans. More employment opportunities were needed. In 2012, the former President Park Chung-hee's daughter, Park Geun-hye was elected president of South Korea. It was long after her father's assassination (and the eight presidents which followed between them). Under the new President Park Geun-hye, the long-standing themes of South Korea's independence and of growth continued.

We recall that back in 1962, General Park Chung-hee had seized control of the South Korean government and, as president from 1963 to 1978, directed the beginning of Korean economic growth. President Park's policies for national wealth and employment facilitated the dramatic rise of South Korea as a global economic power. In 1974, President Park's wife and Geun-hye's mother, Yuk Young-soo, was killed by an assassin. In 1979, Geun-hye's father, President Park Chung-hee, was assassinated.

Park Guen-hye was the first child of Yu Young-soo and Park Chung-hee, born on February 2, 1952. Her younger brother was Park Ji-man and younger sister Park Seoyeong. She attended Jangchung Elementary School in Seoul and Sungshim Girls Middle and High School, graduating in 1970. She entered Sogang University and graduated in January 1974 with a bachelor's degree in electronic engineering, and she went to France to study at the University of Grenoble. When her mother was assassinated in August 1974, Geun-hye returned home to South Korea and assisted her father as official hostess, until his assassination in 1979.

In 1998, Park Geun-hye was elected a Grand National Party (GNP) assemblywoman for Dalseong, Daegu. She retained her constituency seat until 2012, when she resigned it to run for a proportional representative position for the GNP. In 2011, the GNP changed its name to the Saenuri Party (New Frontier). In 2011 Park became chairperson of the Saenuri Party. In 2012, Park ran as presidential candidate and won with 51.6% of the vote, becoming Korea's first woman president.

One of the important issues to President Park Geun-hye was the divide of wealth and employment in South Korea. John Power and In-jeong Choi wrote: "Economic democratization has summed up the political zeitgeist like no other single phrase. President Park Geun-hye latched on to the concept not long into her presidential campaign and subsequently declared it a priority of her administration. While vaguely defined, the idea, when applied to the business realm, implies support for small business in an environment dominated by the colossal chaebols. In a country where a single chaebol, Samsung, generates close to 20 percent of GDP, small and medium business owners have often complained of struggling simply to survive. In response to this perceived unfairness in the marketplace, a mixture of laws and official 'suggestions' have sought to limit the operations of large firms . . . Last month, the government announced a set of policies to help grow small and medium enterprises." (Power and Choi, 2013) For example, the Ministry of Trade, Industry and Energy had decided to triple its research and develop (R&D) funding to SMEs in 2013 and, by 2017, allocate half the R&D budget to SMEs. Politically, the policy of the new Park government appeared much like the policies of the previous Park government – to build a strong and prosperous South Korea, with economic justice for all Koreans.

President Park Geun-hye had three policy themes: science & technology, North Korea, and economic democratization. The first issue of Korean science & technology policy was to facilitate Korea's continuing global economic competitiveness by facilitating chaebol growth in 'creativity'. The second issue errupted in May 2013, when North Korea threatened renewing the war with South Korea and the United States. The third issue addressed Korean businesses smaller than chaebols, often called SMEs. We focus here on the third issue.

Presdent Park Geun-hye. (Source:http://en.wikipedia.org/wiki/Park_Geun-hye)

SMEs and Manufacturing

For SMEs engaged in manufacturing, it is important to review again the concept of an industrial value chain, as SME's principal customers are global companies. We recall that technologies organize an economy into industrial sectors through value-transformations and this was called an *industrial value chain*, sketched again in Figure 10.1.

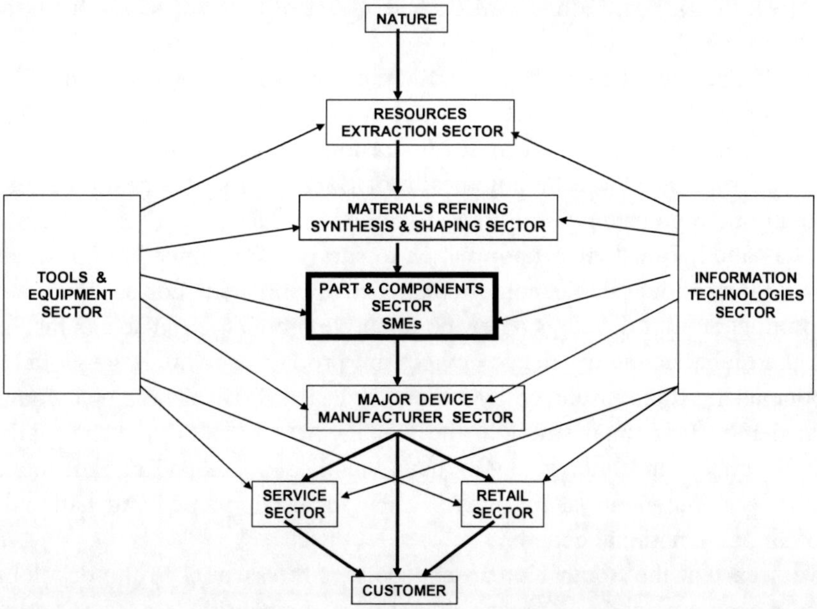

Figure 10.1. Industrial Value Chain.

An industrial value-chain begins with a resource-acquisition industrial sector. The next sector is a materials-processing sector. Resource-acquisition and materials-processing sectors organize separately because the technologies and investments and operations differ. Materials-processing sectors supply materials to part-producing sectors. For example, the automobile industry is supplied by many different kinds of part-producing firms, such as those making tires, wheels, brakes, battery, electronics, etc. The businesses in the parts sector are often the size of SMEs. The next industrial sector is a major-device fabricator sector. For example, in transportation applications, key major-devices are automobiles for land, airplanes for air, and ships for sea. The businesses in the major device sector are large global companies.

Major-device industries or service industries often retail through a distribution system. For example, in the case of cars, there is an automobile-dealer industry for sales and service. In the case of airlines, tickets may be sold directly by airlines or through travel-agents. Finally, the customer is served by the products and services created in the industrial value-chain. These customers may be consumers or business firms (involved themselves in different industrial value-chains) or government. After purchase of the major device, the customer will continue to purchase maintenance, repair, and supply services for the device.

An industrial value chain describes the industrial structures in an economy – connecting industrial sectors which transform resources from nature into products for customers.

The challenge for SMEs in manufacturing is innovating new technologies for quality and cost in the parts sold to the global corporations which produce the major devices.

The customers for an SME manufacturing business are the global corporations which manufacture major devices. This has been called business selling to business, or 'B2B'. For example in 1997, the B2B of Korean SME's had 74.9% of their sales as subcontracts to large companies. (Kim, 1997)

This B2B can be a good business for an SME enterprise – except when the SME has only one big customer and only one product for the customer. Then the SME is economically a captive of its one big customer. If it losses its customer's good will and orders, the SME is immediately bankrupt. For the economic health of B2B SME's, it is desirable for an SME to have more than one big customer and more than one product line. Then if the SME losses a customer, it has another customer, upon which to survive economically until it can gain new customers. And if a product line becomes competitively threatened or obsolete, it can survive on another product, until it develops a new competitive product to replace its lost one.

Thus in the global economy, major devices are produced by the large global companies; while components are often produced by nationally-based SMEs. The global companies have the technical depth, the capital strength, the management breadth, and the market reach to compete in all nations. In contrast, SMEs have technical focus and market niche focus for producing quality components at low cost. Both global companies and national SMEs are important to a modern national economy.

We have seen that the South Korean chaebols were essential to the development of an export-oriented economy, and to Korean international competitiveness. They have created the much of the commercial wealth for South Korea; however, they have not created most of the jobs for South Koreans. The major role in employment has resided with Korean SMEs. South Korea has needed and continues to need both: the global companies (with capital depth and global reach) and the SMEs (for employment and with market niche focus).

German Mittlestand SMEs

Earlier, we noted the success in Germany of SMEs in creating national wealth and providing employment. German SMEs created 51.8% of the GDP, 79.2% of German employment, and hosted 83.2% of apprenticeships, in the German educational/training system for new workers. (Robaschik, 2013) Also we noted that German SMEs operated within a commercial culture unique to Germany, called 'Mittelstand'. The German term 'stand' refers to a status of guild workers in medieval Germany, which handcrafted the goods produced for trade. The term 'mitte' means 'middle', so that mittelstand refers to a middle estate between the military aristocracy and the farming peasants of the upper and lower estates of the feudal German-speaking territories. Guilds were a middle class between the rulers and peasants in European feudal society.

In Germany, as industrialization appeared in the late 1800s, small craft shops still existed down to the First World War. In that nineteenth century, large German firms had been growing and were established in steel (Krupp), in electricity (Siemans), in chemicals (IG Farben), and in pharmaceuticals (Bayer). And these continued after the first war, but the German economy was in turmoil during the 1920s and into the 1930s, due to the penalties and repatriations imposed by the victorious allies of the war and next a global depression. This resulted in Nazi seizure of power and World War II. Also losing the second war, Germany had been devastated and divided into two parts, West and East Germany. Small family owned businesses played a major role in the rapid and successful rebuilding of the industrial economy in West Germany. West German government provided a national partnership, helping the SMEs rebuild Germany. A renewed Mittlestand SME culture was developed in Germany.

In that commercial culture, German Mittlestand companies are focused upon principle product lines and are not run as business conglomerates. They concentrate upon technological innovation to improve quality and lower costs of their specialized products/services. They help in the education and use one of the most highly trained work forces in the world. They use the German university system to acquire and develop new technology. They finance themselves conservatively and employ professional management to assist in the running of their businesses.

Mittlestand firms are particularly export-oriented, aiming to dominate high-quality, technically-advanced niche markets in the global economy. In the European Union (EU) manufacturing industries, German SMEs captured 34% of the EU machinery and equipment business, 28% of the metal products business, and 27% of the rubber & plastic goods business. In 2009, there were 340,000 Mittlestand SMEs exporting goods and services to global markets. Of the manufacturing SMEs, 20% exported goods; 18% SMEs in retail trade exported; 16% of wholesale-trade SMEs exported; 7% of scientific and technical services SMEs exported. To achieve the technical competitiveness for exporting, the German SME Mittlestand business culture was active in Research and Development (R&D). In 2010,

Mittlestand businesses employed 70,000 R&D personnel and spent 9% of their budgets on R&D. (Robaschik, 2013)

The German Federal Ministry of Economics and Technology cited the following reasons for the success of the German Mittelstand:

> "Almost all of Germany's SMEs are family-owned, and many are managed by their owner. Their business policies tend to be especially long-term.
>
> "The German Mittelstand companies are some of the most innovative in Europe; 54% launched an innovation onto the market in the 2008-2009.
>
> "The German Mittelstand relies on sound financing models – chiefly equity and bank loans.
>
> "The high equity ratio and a cautious approach to expansion enable the companies to undertake medium-term and long-term investments, even in times of crisis.
>
> "The German Mittelstand employs 83.2% of trainees, the much sought-after skilled workers of tomorrow.
>
> "Government supports the German Mittlestand on key issues like: investment in R&D, the skills shortage, foreign trade and investment, financing needs, company start-ups and company hand-overs." (Federal Ministry of Economics and Technology, 2013)

National Innovation System

Also as we have discussed in the earlier case of Nano, the first important challenge for long-term economic survival of a manufacturing SME is to acquire new technology upon which to build a new product line. The problem is that SME's seldom have the science base to create new technology. In a modern nation, the science base in a society resides in the research university infrastructure of its economy, and this had been called a 'national innovation system". And we recall that in the case study of Nano, we saw university research go directly into a new high-tech SME. But how can an existing SME obtain new technology from universities, when the owner is not a university professor? This has a problem called 'technology transfer' from universities to industry. In a modern nation, research and development (R&D) is organized as a research infrastructure – with universities performing scientific research, government funding university scientific research, and industry performing technology research and development and product innovation. The innovation processes in the system are the procedures by means of which the three sectors interact for technological innovation.

For example, in the United States the biotechnology industry began on the basis of a university research project, a recombinant DNA experiment by Professor Stanley Cohen at Stanford University and Professor Herbert Boyer at University of California, San Francisco. The research was funded in part by a government agency, National Institute of Health, and which continued to university research on genetics. The first biotech company, Genentech, was started by Boyer and Robert Swanson, a venture capitalist. Stanley Cohen became a research director in another early biotech company Cetus. We sketch this as a set of institutional areas (Venn diagrams) organized for the transformation of knowledge from nature to utility, as shown in Figure 10.2.

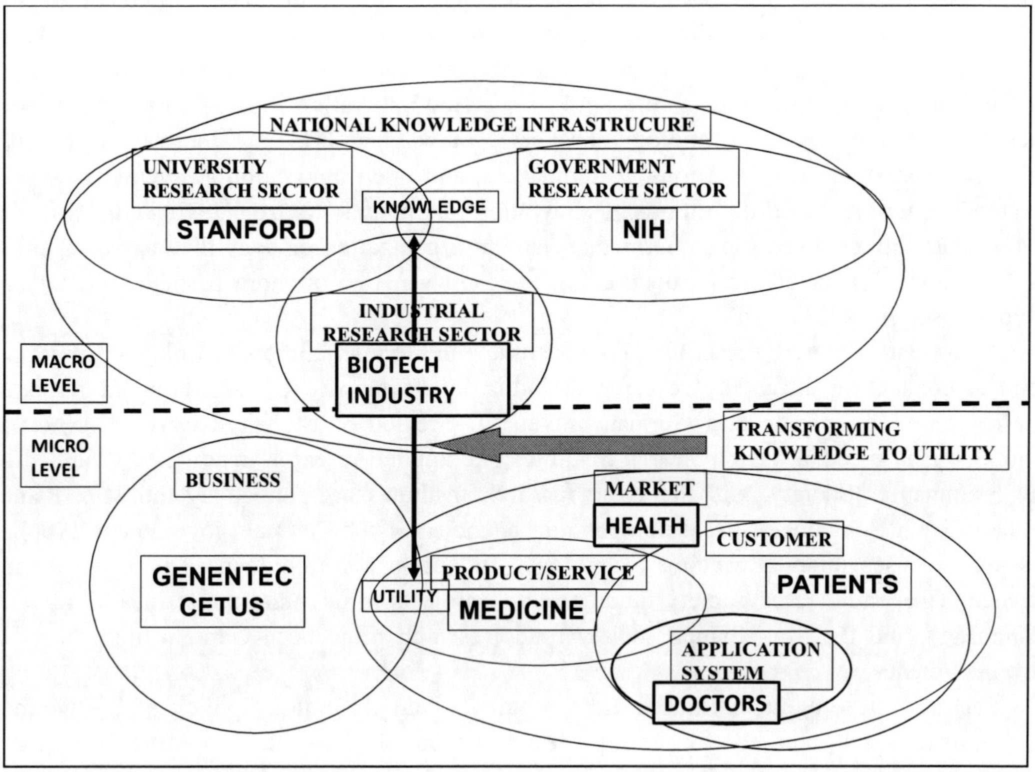

Figure 10.2. National Innovation System Institutionalizing the Transformation of Knowledge to Value.

For large corporations with a corporate R&D lab, the researchers in the corporation can access university science research directly. Scientists employed in corporate R&D can communicate directly with university scientists. Thus an industrial intermediate connection is usually not necessary for global companies. This is the basic technology challenge to an SME's competitiveness.

For an SME to sell its products to a global company (as a source for components), the SME must design and produce its products at the cutting-edge of technology – high tech quality and low cost.

But since an SME cannot afford an advanced R&D lab (necessary for the cutting-edge of technology), where can an SME get advanced technology? The answer is from research universities in a national research system. This raises the important issue of how to transfer university research into SME technology? Thus for SME's, a technology transfer intermediary organization is needed, to get the university research into new technology and the new products/services of an SME. Germany, with its Mittlestand business culture has been one country very successful in technology transfer from universities, We next look at Germany's technology transfer infrastructure, as their program of the Fraunhofer Society.

German Fraunhaufer as an Innovation Intermediary

Antonie J. Jetter and Songphon Munkongsujarit have discussed the concept of an intermediary organization to facilitate the technology transfer from science and technology at a national level down into the technology acquisition and development at a firm level, operating as an *Innovation Intermediary*. They wrote: "There are a number of different terms in the literature that are used to refer and to describe innovation intermediary. These terms include: intermediaries, technology brokers, knowledge brokers, innovation brokers, bricoleurs, boundary organizations, matchmakers, and open innovation accelerators. One of the most commonly used definitions of innovation intermediary is from J. Howells [Howells, 2006] who defines innovation intermediary as 'an organization or body that acts as agent or broker in any aspect of the innovation process between two or more parties'." (Jetter and Munkongsujarit, 2013)

As we earlier noted, the Fraunhaufer institutes in German universities play an important role, in connecting university research to industrial development. Each Fraunhaufer is an applied research institute at a German university, directed by a chaired German professor, who also may direct a basic research institute. The applied research institutes are funded by the Fraunhaufer Society, and the basic research institutes are funded by the Max Planck Society. Both Societies are research funding agencies of the German government. Applied research focuses upon advancing technology, and basic research focuses upon advancing science. The Fraunhaufer Society funds research projects in German universities to advance technology; and the Max Planck society funds research projects in German universities to advance science.

Together these agencies implement the science and technology policy of the German government. Each chaired German professor in engineering may direct two research institutes, applied and basic. The Fraunhaufer Institute of a German professor is focused in advancing technology by connected the basic research of science to the applied research of technology. However in industry, applied research must next be embedded in a product (or process or service) development project. It is the sales of products and services that earn industrial income (not the sales of technology).

The intermediary innovation role of the Fraunhaufer institutes is to help translate technology progress into product development – new products (or processes or services) for businesses. This translation requires applied research to adapt the the technology to a product requirement and design of a new product which embeds the new technology. The commercial advantage of this is a new high-tech product/service, which provides a competitive advantage of higher quality and/or performance than existing products.

Fraunhaufers focus upon product and production development. To assist in introducing new prototype production processes, they train a business's staff on site and assist with marketing plans. Antonie J. Jetter wrote: "Fraunhofer is an extremely valuable resource for industry, especially for mid-size companies. Fraunhofer shapes industries through services and its influence on government research programs. Fraunhofer tends to make strong industries stronger." (Jetter, 2011)

In 2010, Germany had 80 Fraunhaufer institutes, with 18,000 technically qualified employees, and an annual budget about 1.65 billion euro. This budget came from about one third base-funding from the government Fraunhaufer society, one third from project contracts, and one third from industrial contracts. The project funding came from German federal and state sources (e.g. Ministry of Science and Technology, Ministry of Economics and Employment, etc.). Projects arise from proposals which an industrial and university partner defines and writes a research and development (R&D) proposal and applies to a government agency for its funding. (Jetter, 2011)

An example of one of these institutes was the Fraunhofer Institute for Production Technology (ITP) and the Laboratory for Machine Tools and Production Engineering (WZL) at Aachen University. These two facilities cooperated for projects in manufacturing research and development and were located in the Faculty of Mechanical Engineering at Aachen University (which held 50 professors and 6700 students and 2200 employees). The budget in 2011 for both ITP and WZL was 32 million euro, of which 40% was from public funds and 37% from industrial projects, with 23% base funding from the Fraunhofer society and Aachen University. (Jetter, 2011)

ITP focused research on process technology, production machines, metrology and quality management, and technology management and purchasing. Professor Kloche directed research in process technology; Professor Brecher directed research in production machines, Professor Schmitt directed research in Metrology and quality management; Professor Schuh directed research in technology management and purchasing; Professor Sharon directed research in manufacturing innovation.

WZL focused research upon manufacturing technology, machine tools, metrology and quality management, and production engineering. Professors Klocke directed the research in manufacturing technology; Professors Kocke and Brecher directed the research in gearing technology; Professor Brecher directed the research in machine tools; Professor Schmitt directed the research in metrology and quality management; Professor Schuh directed the research in production engineering.

In process technology, research was then focused upon machining and material removal processes, laser materials processing, forming processes. Production and machine tool research focused upon design of machines, control and testing. Gearing technology focused upon gear manufacturing, gear calculation.

Management research focused upon business development techniques, technology management techniques, innovation management techniques, production and quality management techniques.

Thus the Aachen Fraunhofer institute was organized in two facilities specializing in production, ITP and WZL0. These focused upon manufacturing technology and were directed by five chaired professors – all in the mechanical engineering faculty of Aachen University (Reinisch-Wesfalishe Technische Hoschshule Aachen). One can see that the Fraunhofers specialize in certain industrial technology areas, performed research in these areas, from technology development to new product-process-service design and implementation.

ADAPTING MITTLESTAND INTO KOREAN BUSINESS CULTURE

German Mittlestand development has successful in helping Germany create employment and exports for German SMEs. If a Mittlestand model were to be adopted by South Korea, what kinds of kinds of issues would the Korean SMEs and Korean government have to solve in order to adapt Korean business culture to a Korean Mittlestand culture?

1. Issues in the Acquisition of New Technology for New Product Development in SMEs

University-based research is an important national resource for SMEs to acquire new technologies to develop competitive new products and services, but difficult to match to SME needs and capabilities.

2. Issues in Government Policy for Creativity in National Mittlestand Infrastructures

Government needs to provide research grants for the funding of innovation intermediary activities between universities and SMEs. For this, appropriate government policies are needed to facilitate the research alliances between universities and SMEs and global corporations – to facilitate the growth of employment opportunities in a national economies.

3. International research on 'Windows on Science' for SME and Global company strategic alliances

Innovation proceeds not only in incremental steps but also by the discontinuities of next generation technologies. Incremental innovation can arise from applied research; but discontinuous innovations arise from basic research. Strategic research alliances between SMEs and large global companies can occur around and in selective areas of scientific advance – mediated by Fraunhaufer innovation activities. In the United States, the Engineering Research Program of the National Science Foundation provides funding for discontinuous innovation partnerships between universities and industry.

4. Issues on Venture Capital and SMEs

New high-tech products or services innovated by SMEs through university-connected Fraunhaufer institutes can provide opportunities for business expansion by SMEs – which mary require financing. This provides opportunities for venture capital in a nation to assist in the financing of SME expansions. Appropriate partnership arrangements are needed for the interests both of SMEs and venture capital firms.

5. Issues for SME Governance

SMEs are predominantly family enterprises. As a family generation replaces the founding generation, issues about management and governance arise. Also the expansion of SME businesses through additional product lines raises management challenges. In the German Mittlestandt experience, it has been important to replace family management with professional management. There are governance challenges in this transition, adding professional skill while still maintaining appropriate family control.

LESSONS LEARNED

Germany's Mittlestand SMEs are family owned and also are technology leaders in their niche markets. We saw that Nano had a potential of providing a model in a Korean Mittlestand. It was founded with a kind of 'family' orientation to provide employment for a professor's graduates. It was a technology leader in the invention of a new process to produce Nano-sized catalysts. It was unique educational program lasting more than 10 years for the employee to improve not only their technological competence but also management strength. The difference of Nano from traditional Korean SMEs was that the technology of Nano came directly from the research of a Korean university professor. But in the German Mittlestand context, there are German government-sponsored organizations established to assist Mittlestand companies in acquiring new technology (e.g., Fraunhofer-Gesellschaft and Steinbeis-Stiftung, and Gründer-Technologie-Transferzentren). Similar national infrastructure aspects for technology assistance to SMEs need yet to be constructed in the Korean national innovation system.

Also an importance difference did appear in the application of the theory of industrial value-chains. Manufacturing SMEs are often/usually suppliers to other industrial firms, in what is called, Business-To-Business (B2B). For example, in the German Mittlestand SMEs "about 90 % of the 1500 German world market leaders are active in the B2B segment". (Venohr, 2010) So too Nano's customers were also as B2B (Bundang, SK, EnBW and other power plants in the world). What was different was that Nano was begun with an additional strategic alliance relationship to the Korean chaelbol, SK. Nano was not merely a supplier to SK but also a strategic technology partner. The professor founding Nano had invented and patented the innovative technology of TiO_2 nano-particle. SK initially contracted with Nano to use powder for SK's customer of EnBW. SK then encouraged Nano to continue to develop the product capability of supplying honeycomb catalyst. SK thought it not a large enough market for SK to invest in developing its own production capabilities.

Adaptation of a 'Mittlestand' model for SMEs in a nation requires adjustment to the culture and national research infrastructure of the nation.

Chapter 11

INDUSTRIALIZATION – ECONOMY AND SOCIETY

INTRODUCTION

We have seen that economic theory, of one form or another, has been used to formulate government policy for economic development. President Park used a development theory of 'export economy'. IMF officials used an economic theory of 'balance of imports-exports' as a policy for economic development.

The difference was that Park's government emphasized technology acquisition and new company growth; whereas IMF officials emphasized increased interest rates and increased unemployment. Each policy was justified by different economic theories. Park's economic theory was non-traditional economic theory; whereas IMF's theory was traditional economic theory, called Laissez-Faire theory.

Laissez-Faire economic theory has been the dominant (mainstream) theory of many of the economists in the world. How did this come about? 'Laissez-faire' is a French term for 'let them do' and has been used to label a free market economy – an economy where government lets business do exactly 'what they will' (laissez faire). Government leaves the free-market alone, without regulation. The advantage of a 'free-market economy' is that supply-and-demand in the market will set optimal prices for commodities in the market. Historically, governments which interfered by controlling prices frequently induced shortages-of-commodities, inflation-of-prices, and black-markets. Modern economic theory is focused upon markets with freely competing producers and their competition setting prices – as the most efficient economy. Economic history has provided abundant evidence that laissez-faire works. For example, we have seen in the differences between South Korea's prosperous economy and North Korea's starvation economy. However, Laissez-Faire is an incomplete theory of an economy. It focuses only upon how businesses should operate but does not depict the role of government in proper regulation of a market. A Laissez-Faire market without proper regulation can have fraudulent, unsafe products and be unstable, as the world recently witnessed in the 2007-08 Global Financial Crisis.

Leading up to that crisis, 'Laissez-Faire Theory' was used not just as economic theory but as political ideology – ideology for no regulation of derivatives and little regulation of banks. Ideology that all government regulation is a hindrance on the free market activities of business. And all free markets operate perfectly. Central bankers in the U.S. and Great Britain used the ideology to avoid regulation of financial markets. In 2000 in the US, Allan

Greenspan had been Chair of the U.S. Federal Reserve and opposed government regulation of financial markets. Paul Goodman wrote: "Mr. Greenspan said that Wall Street could be trusted. He said, 'There is a very fundamental trade-off of what type of economy you wish to have. You can have huge amounts of regulation, and I will guarantee nothing will go wrong, but nothing will go right either.' (Goodman, 2008) But in 2008, harsh empirical economic reality recurred and was witnessed, even by Alan Greenspan. Paul Krugman wrote: "By October of last year, however, Greenspan was admitting that he was in a state of 'shocked disbelief,' because 'the whole intellectual edifice' had collapsed. Since this collapse of the intellectual edifice was also a collapse of real-world markets, the result was a severe recession — the worst, by many measures, since the Great Depression. What should policy makers do? Unfortunately, macroeconomics, which should have been providing clear guidance about how to address the slumping economy, was in its own state of disarray." (Krugman, 2009)

How did a theory of 'Laissez-Faire economics' become a political ideology of 'Laissez-Faire deregulation'? Why was a reasonable economic theory misused as an 'economic ideology' – a justification for government not doing its proper regulatory job? Misuse of theory has had damaging consequences to the world – to the U.S. in 2007-08 and to Asian countries in 1997-98. How could some economists and regulators forget that a real free market must be properly regulated by government – for honesty, safety, stability? To see how *economic theory* became confused with *economic ideology*, we briefly review how economic theory arose.

Laissez-Faire economics has provided the reasoning and basic concepts about industrial civilization – and has been historically linked to societal upheavals which have accompanied the industrial revolution. In the world, the industrial revolution has been occurring over centuries, over the 1800s and 1900s, as a tide of social change rising in the world – and upon which the new South Korea rose as a globally competitive nation.

The basic economic fact of industrialization is this. *Mechanized industrial technology is so productive, it produces many more commodities than can be sold locally – and so products must be exported.* Industrialization impacted international relations between nations by increasing dramatically the need for international trade – production exports and resource imports. For example, South Korea must export manufactured products, such as automobiles and IC chips, to gain national revenue to import oil resources (since South Korea has few natural oil resources).

Over the three hundred years from 1700 to 2000, the world changed from thinking about a nation as an international empire (e.g. the British Empire or the Japanese Empire) to a nation of global corporations (e.g. Korean chaebols or Japanese keiretsu or U.S. and British global corporations). This 'national' concept evolved in the 1800s and 1900s, but in problematic ways. What is the social basis of a 'nation'? How does an 'industrial economy' work? What is 'a national people' (what the Germans called 'ein Volk')? What are the rights of the 'state' versus those of the 'individual'? The societal struggles for the answers to these ideas have been the principle themes. Coincidentally as the social sciences emerged, the disciplines of economics, sociology, political science, and anthropology have studied these same themes. Not only did society change in the industrial revolution, but the thinking-about-society has focused on the basic concepts of 'nation', 'industrialization', 'civilization'.

Ideas about a nation, about of society, about the state – all were central to the violent wars of the twentieth century. The First World War was triggered by 'Serbian' nationalists assassinating a future emperor of the Austro-Hungarian Empire. Germans, French, British,

Russian, Italian peoples went to war for their nations. The Second World War was waged by nations as an axis of fascist dictatorships versus allies of democracies (with the exception of the communist Soviet Union). And the allied victory over fascism next devolved into a 'Cold War' between capitalist democracy and communist dictatorships. This chilling episode of a cold war never erupted into a hot' war, fortunately. The atomic bomb had invented in the Second World War, and any hot war thereafter portended the destruction of everything in a nuclear Armageddon.

In the rise of the global civilization (all under the threat of a nuclear cloud), South Korea found its opportunity to industrialize for freedom. We have seen that a proper kind of partnership between industry and government was essential in that industrialization. Lacking such a proper partnership, North Korea stagnated into a failed dictatorship state. What is the philosophical basis for a 'proper partnership' between industry and government, to attain national freedom and prosperity?

History: Technology and Industrial Revolutions

We should briefly review the historical interaction between the new technologies and economic growth in the industrial revolution. The basic interaction of the economy and technological progress and industrial expansion was first documented by a Russian economist, Nicoli Kondratiev. In the 1920s, he studied the economic development in the English economy of the late 1800s and early 1900s (and later was imprisoned and shot on Stalin's orders). In the 1930s, another economist in Austria, Joseph Schumpeter, read and shared Kondratiev's idea – that innovation was important to economic growth. (Schumpeter, 1942) George F. Ray wrote: "...the work of Schumpeter...put emphasis on innovation and on the subsequent burst of entrepreneurial investment activity as the engine in the upswing of the long cycle, a la Kondratiev...." (Ray, 1980) In 1990, Robert Ayers brought up to date the empirical correlation between European industrial expansion and contraction and the occurrence of new technology based industries (Ayers, 1990).

1770-1825 Economic Expansion and Contraction (1826-1847)

The beginning of the industrial revolution in Europe was based on the new technologies of steam power, coal-fired steel and textile machinery.

The English economy expanded from 1792-1825, with steel production, steam engines, and textile machinery providing the expanding industries of the economic expansion. But as manufacturers expanded production capacity, temporary excess capacity triggered an economic contraction from 1826-1847.

The invention of the steam-powered engine had required the science base of the physics of gases and liquids. This new scientific discipline of physics provided the knowledge base for Thomas Newcomen's and James Watt's inventions of the steam engine. Coal-fired steel required knowledge of chemical elements from the new science base of chemistry which was to be developed in the middle of the 1700s The new disciplines of physics and chemistry were necessary for the technological bases of the industries of the first wave of economic expansion from the beginning of the industrial revolution in England.

1847-1873 Economic Expansion and Contraction (1873-1893)

The acceleration of the European industrial revolution was based on the technologies of railroads, steamships, telegraph and coal-produced gas lighting. There was a second period of economic expansion in England from 1847-1873, followed by a contraction from 1873-1893. The expansion was due to the beginnings of new industries in railroads, steamships, telegraph, and goal gas but again temporary excess capacity triggered the following economic contraction. The expansion to which the telegraph contributed was based on new scientific discoveries in electricity and magnetism in the late 1700s and early 1800s.

1895-1930 Economic Expansion and World War I (1914-1918)

In the third long economic expansion, it was again the new physics of electricity and magnetism that provided the science bases for inventions of electrical light and power and the telephone. Also advances in the new discipline of chemistry provided the science base for the invention of the chemical dyes. Artificial dyes were in great economic demand because of the expansion of the new textile industry. Producing these dyes added to the production of gun powder, so that industry expanded into the modern chemical industry. The economic expansion was also fueled by the invention of the automobile, which depended upon the earlier invention of the internal combustion engine. This internal-combustion engine invention required knowledge bases from chemistry and physics. Radio was another new invention based upon the advancing science in physics of electricity and magnetism. Chemical plastics were invented by further experimentation in the advancing scientific discipline of chemistry.

The First World War ended the economic expansion in Europe, when production was turned to military weapons and munitions. After the war, Russia was taken over by Bolshevik communists, Germany labored under war indemnities, eventually turning into a Nazi dictatorship. England and France slowly recovered economically In the US., economic expansion continued after the war, driven principally by the new technologies of the automobile and airplane and radio. But excessive financial speculation brought down the U.S. economy in the 1929 New York stock market crash. The U.S. entered a decade of depression, which spread around the world, contributing to the emergence of the dictatorships, fascist in Italy and Germany, and military in Japan.

Economic Long Wave

This interaction between technological progress and industrial expansion-contraction is called a 'Kontratiev Economic Long Wave'. The contractions of economic activity followed after the expansions of economic activities due to the innovation of new technologies. Why should economies first expand and then later contract after technological innovation?

Historically in the late 1800s and early 1900s, why didn't the English or other European economies continue to grow smoothly as the new technologies created new industries? Kondratiev observed both expansions and contractions in the English economy. Why always was an expansion followed by a contraction? The answer lay in the interaction between technology and finance.

At first, technological innovation attracts capital investment which produces returns on investment as new high-technology products and services are sold into an expanding market. But this business expansion attracts competition. Further financial investment launches competing businesses in the new market for high-technology products and services. And even while the new high-technology based industry markets grow, industrial production can grow even faster. This leads to a temporary excess production-capacity, which results in price competition and some business failures. Business expansion can be followed by business contraction when financial investment creates temporary excess supply to demand. The stages in an Economic Long Wave consist of:

1. Science discovers phenomena which can provide for a new manipulation of nature by technological invention.
2. New basic technology provides business opportunities for new industries.
3. A new high-tech industry provides rapid market expansion and economic growth.
4. As the new industry continues to improve the technology, products are improved and prices decline and the market grows toward large volume (creating the previously discussed 'Industrial Life Cycle').
5. Competitors enter the growing large market, investing in more production capacity.
6. As the technology begins to mature, production capacity begins to exceed market demand, triggering price cuts.
7. Excess capacity and lower prices cut margins and increase business failures and raise unemployment.
8. Turmoil in financial markets may turn recession into depression.
9. New science and new basic technologies may provide the basis for a new economic expansion.

Thus the interaction between technology and economic expansion is mediated by finance – by the financial system. But how does this interaction also impact societal structure? Since capitalism flourishes under democracy and is stifled or corrupted by dictatorship, why did capitalism in the industrial revolution not automatically lead not to democracy? As Kondratiev had explored the connection of technology to economic expansion, but the other connection needed to be understood is that of technological industrialization to societal structure.

Society and Laissez-Faire Economies

This technological-change-of-economics impacted how European society thought about itself. How did it change the European peoples' concepts of 'nations' and of 'culture'? One economist deeply interested in this question was Karl Polyani. In the midst of the Second

World War in 1944, Polyani wrote about the cultural puzzle of fascism and communism and of democracy. How had the European ideas of the nineteenth century about their nations as 'empires' (e.g. British Empire, French Empire, German Empire, Russian Empire) change to the twentieth century ideas of communist and fascist ideologies (e.g. the Soviet Union or Fascist Italy or the Nazi Reich)?

Karl Polyani wrote: "Nineteenth-century civilization has collapsed. . . . (It) rested upon four institutions. The *first* was the balance-of-power system which for a century prevented the occurrence of any long and devastating war between the Great Powers, after 1812 and up to 1914. The *second* was the international gold standard which symbolized a unique organization of world economy (in the 1800s). The *third* was the self-regulating market which produced an unheard-of-material welfare. The *fourth* was the liberal state. . . . Two of these institutions were economic, and two political; and . .. two of them were national and two international." (Polyani, 1944)

What is relevant is that these factors still are current. The twenty-first century civilization of the global market and corporation may be as unstable, as was the twentieth century of international civilization. For example, in the first two decades of the twenty-first century, two financial crises have already occurred: the global financial collapse of 2007-2008 and the Euro-dollar crisis of 2010-2013. We compare the relevance of Polyani's focus on world 'institutions'

> *First*, there are still 'Great Powers' in the world of global civilization, particularly the United States and China (now called 'Super Powers'). And we saw the meeting of their leaders in 2013, as a response to the North Korea crisis of March 2013 – a modern kind of' balance-of-powers' to deal with North Korean nuclear threats. But can there be any kind of 'balance-of-power' in the twenty-first century, when all-out-war with nuclear weapons is suicidal and yet it could be triggered by a single small country launching a nuclear missile?
>
> *Second*, there remains the need for an international currency standard. There no longer is a gold standard, since its financial role was replaced the U.S. dollar – until the U.S.-induced financial collapse of securitized investments in 2007. But the need for an international monetary standard still exists, as evidenced by the failure as yet of the Euro-dollar to play an international role in finance. Also the Chinese monetary unit cannot provide an international monetary role (as long as the government is still one-party communist and its economy socialist-capitalistic).
>
> *Third*, the self-regulating international market has been empirically shown that it requires international financial regulation in financial markets for the safety of central banks.
>
> *Fourth*, the idea of a 'liberal state' is still being puzzled out – particularly about how much a welfare structure a government can provide for science, education, health, pensions, energy and – how financed by taxes. Who pays taxes and how much – individuals or corporations? How can a liberal state ensure both full-employment and economic competitiveness?

The basic themes, in the questions Polyani's posed about the collapse of nineteenth-century international civilization collapsed, are still with us. What is and should be an industrialized society? Key concepts of an 'industrialized society' are about 'what-is' and 'what-ought-to-be', as these are the empirical and normative depictions of an industrial society. Is an industrial society merely a Laissez-faire economics free market economy? Or is it more? Polyani argued that the ideas of 'democracy' and 'capitalism' became intertwined, so that in the twentieth society the two concepts were inseparable. But their connections were problematicr. One extreme idea of a nation was that economics can totally dominate politics, – economic liberalism. The opposite extreme idea of a nation was that politics can totally dominate economics – fascism and communism.

But what we have seen in the case of the two Koreas is that neither concept (economic liberalism nor communism) wholly characterized the history of the two Koreas. North Korea passed from a communist dictatorship to a military dictatorship (Kim Jung-il and Kim Jong-un's military-first policy). South Korea is a nation in which economics and politics achieved a balance, in a patriotic partnership.

How did these two extreme views on society (either a nation of economic liberalism or a nation of fascism/communism) arise in history, from the industrial revolution of the nineteenth century? This was Polyani's question.

Karl Polyani first focused on the idea of 'economic liberalism', which is the view of Laissez-faire economics (free market) as the *whole* of an industrialized society. How did this model of a modern industrialized economy emerge as the idea for 'utopian' industrialized society? How did it become the political ideal of capitalistic societies in the international civilization of Europe in the 1800s to 1900s?

Polyani agreed that free markets and free trade and capitalism were essential ingredients in a modern industrial society – but not the whole of all society! And that's why he called 'economic liberalism' a 'utopian' view of society. Polyani wrote: "Economic liberalism was the organizing principle of society engaged in creating a market system. . . . Only by the 1830s did it stand for the three classical tenets: that labor should find its price on the market; that the creation of money should be subject to an automatic mechanism; that goods should be free to flow from country to country without hindrance of preference; in short, for a labor market, the gold standard, and free trade. . . In the 1830s, economic liberalism burst forth as a crusading passion and Laissez-faire economics became a militant creed." (Polyani, 1944)

Polyani saw 'classical economic theory' as a set of 'principles' – necessary but not sufficient for an industrialized economy to function. Sufficiency required some government context and regulation. Otherwise, an industrial society organized only as Laissez-faire economy would be unstable. Yet classical economic theory ignored the government role in stabilizing a free-market economy. Some classical economists worried that any government interference would impede the pricing mechanism in a free market. No regulation! Deregulation! (This was political position the Chair of the U.S. Federal Reserve System Allan Greenspan from 2000 up to the U.S. financial collapse in 2008.)

Polyani had no objection to 'free market' principles. He saw this as correct – as generalization of how a properly functioning economic system (of mechanized production traded in free markets) should work. But he saw also that such economic theory completely subordinated principles of the political system. Applying the principles-of-the-economic-system over principles-of-the-political-system was the 'utopianism' in 'economic liberalism'. In nineteenth century Europe, economic theory became political ideology – an ideology arguing for a 'lasses faire' market – over any other needs and despite any other needs in society.

As an economic historian, Polyani traced the origin of the concept of economic liberalism (of a 'Laissez-faire economics' economic and political system) to a historical context of ideas, arising during the English society's transition from an agricultural-trading society to an industrial-trading society. Polyani's critique was that a 'Laissez-faire economics' economic liberalism (free market system) should have been only a model of the economic system in a society but became a model of the whole English society. Polyani wrote: "Nothing could seem more inept than the attempt to reduce a civilization, its substance and themes, to a hard-and-fast number of institutions; to select one of them (economic system) as fundamental and proceed to argue the inevitable self-destruction of civilization on account of some technical quality of its economic organization. Civilizations, like life itself, spring from the interaction of a great number of independent factors, which are not, as a rule, reducible to circumscribed institutions." (Polyani, 1944)

For Polyani it was this 'idea' (to reduce a whole society to a merely an economic system) which helped set the stage for the subsequent failure of European civilization. Polyani considered World War I and then World War II as failures of European society. Not just unfortunate events, but failures of a social order. The failed social order was that of industrialized nations bound only in economic 'trading relations'. Polyani wrote: "If the breakdown of our civilization was timed by the failure of world economy, it was certainly not caused by it. Its origins lay more than a hundred years back in that social and technological upheaval from which the idea of a self-regulating market system sprang in Western Europe. . . After 1815, the change (in Europe) is sudden and complete. The backwash of the French Revolution reinforced the rising tide of the Industrial Revolution in establishing peaceful business as a universal interest." (Polyani, 1944) What Polyani set out to trace was how, for policy, the concept of a whole society could be reduced to merely an economic system.

A modern societal system contains not only an economic system but also a political system and a technological system and a cultural system. We recall that one can model an industrialized society as a set of such systems, Figure 11.1.

	UTILITY	*IDENTITY*
RECIPROCITY	ECONOMIC INFRASTRUCTURES & PROCESSES (ECONOMIC SYSTEM)	CULTURAL INFRASTRUCTURES & PROCESSES (CULTURAL SYSTEM)
AUTHORITY	TECHNOLOGICAL INFRASTRUCTURS & PROCESSES (SCIENCE & TECHNOLOGY SYSTEM)	POLITICAL INFRASTRUCTURES & PROCESSES (POLITICAL SYSTEM)

Figure 11.1. Taxonomy of Societal System.

Thus Polyani's critique is poignant, his critique of elevating the principles of only the economic system onto the ideology of all society. This ignored the societal needs in culture, technology, and politics. This ignoring of other system needs in a society distorts the society, unbalances a society - leading to tragic events, World Wars I and II in the twentieth century.

Also in the beginning of the twenty-first century, there have been other societal failures – the Asian financial crisis in 1998, the Global Financial Crisis in 2008, the European Euro Crisis in 2010. The long history of bank panics show the need for a complementary political system of financial regulation to enable financial markets to operate as a 'free market'. Laissez-faire financial markets do require proper government regulation, for both the integrity and the stability of a financial market.

RELATION OF ECONOMIC AND POLITICAL SYSTEMS

Polyani's concern was about a proper relation between industry and government, a relationship which had not been properly addressed in the ideology of 'economic liberalism'. Polyani wrote: "The nineteenth century . . . was England's century. The industrial Revolution was an English event. Market economy, free trade, and the gold standard were English inventions. . . At the heart of the Industrial Revolution of the eighteenth century, there was an almost miraculous improvement in the tools of production, which was accompanied by a catastrophic dislocation of the lives of the common people. . Economic liberalism misread the history of the industrial Revolution because it insisted on judging social events from (only) an economic viewpoint. . . (And) the story has been told innumerable times: how the expansion of markets, the presence of coal and iron as well as a humid climate favorable to the cotton industry, the multitude of people disposed by the new eighteenth-century enclosures, the existence of free institutions, the invention of the machines, and other causes interacted in such a manner as to bring about the Industrial Revolution." (Polyani, 1944)

At the heart of the industrial revolution were the new machines, powered by the steam engine, to vastly increase production. In the 1700s, the steam engine was invented to power production machinery and new inventions of textile machine for combing cotton, spinning thread, and weaving cloth – all together they innovated the new manufacturing factories. But the machinery was expensive, requiring a managerial strategy to optimize their profitability. Poyani wrote: "Since elaborate machines are expensive, they do not pay unless large amounts of goods are produced. . . For the merchant this means that all factors involved must be on sale. . . What he buys is raw materials and labor – nature and man. Machine production in a commercial society involves no less a transformation than that of the natural and human substance of society into commodities. . . . the dislocation caused by such devices (machine production) must disjoint man's relationships and threaten his natural habitat with annihilation." (Polyani, 1944)

This change in the organization of production through machinery required that the whole economic system of society be altered. Polyani wrote: "Market economy implies a self-regulating system of markets; it is an economy directed by market prices and nothing but market prices. . . No society could naturally live for any length of time unless it possessed an economy of some sort; but previously to our time no economy has ever existed that, even in principle, was controlled by markets." (Polyani, 1944) *A market economy required by an industrial revolution changes society, as it did in England and Europe in the 1800s.*

Previously in non-industrialized society, economic order is controlled not by an industrial entrepreneur but by a tribal or feudal structure. Polyani wrote: "(In primitive societies) how is order in production and distribution ensured? The answer is provided by two principles of behavior not primarily associated with economics: *reciprocity and redistribution*. The individual's economic interest is rarely paramount, of the community keeps all its members from starving, unless it is itself borne down by catastrophe, . . . The maintenance of social ties is crucial. First, because by disregarding the accepted code of honor . . the individual cuts himself off from the community and becomes an outcast. Second, because all social obligations are reciprocal, their fulfillment serves also the individuals' given-and-take' interests best. . ." (Polyani, 1944) In addition to the principle of 'reciprocity' in non-industrial society, a second principle of 'redistribution' of goods smoothes some of the unevenness of wealth in a tribe. Polyani wrote: "The principle of redistribution . . . (is that) a substantial part of all produce . . is delivered . .. to the chief who keeps it in storage. . . The members of a hunting tribe usually deliver the game to the headman for redistribution. . . All large-scale (ancient) economies were run with the help of the principle of redistribution. . . The redistribution helps to enmesh the economic system proper in social relationships. . . ." (Poyani, 1944) *In non-industrialized societies, the accumulation of wealth is not the central focus of individuals nor the community.*

However, in an industrialized society, wealth becomes the focus, and 'equilibrium price' control provides the ideal operation of the economic system. Polyani wrote: "A market economy is an economic system controlled, regulated and directed by market prices; order in the production and distribution of goods is entrusted to this self-regulating mechanism. . . . self-regulation by 'sales. Accordingly, there are markets for all elements of industry, not only for goods but also labor, land and money. (Their prices called respectively: wages, rent, and interest.)" (Polyani, 1944) *Thus to establish a market economy, not only must the produced goods be priced by a market (commodity prices) but also must the costs of the production be priced.*

Products are only the output of production. Natural resources, labor, and finance, land (location and transportation) are the inputs to production. But these too must be priced. Polyani wrote: "The crucial point is this: labor, land, and money are essential elements of industry; they also must be organized in markets . . . . But labor, land, and money are obviously not commodities; and the postulate that things bought and sold must have been produced for sale is emphatically untrue. . . Nevertheless, it is with the help of this 'fiction' (that labor, land, money are commodities) that the actual markets . . . organized. Labor, land, money are sold in 'markets'; . . . and any measures or policies that would inhibit the formation of such markets would *ipso facto* endanger the self-regulation of the system." (Polyani, 1944) *The costs of production lie in the prices of natural resources, labor, money, and land.*

In a Laissez-faire economy, prices are set for produced commodities. But prices must also be set for inputs to production (labor, resources, finance, land), even though such industrial inputs are not produced as commodities. Labor is not produced; it is born in biological families. Natural resources are not produced; they are mined or grown from nature. Finance is not produced; it is accepted as a medium of economic transactions, money and credit. Again Polyani has no disagreement with classical economics about pricing, pricing of commodities, labor, natural resources, finance. But Polyani thought it important for societal theory to be clear that only commodity goods are produced as outputs in an industrialized economy, while labor, natural resources, and finance may be priced but they are not industrially produced. From whence does labor, natural resources and finance arise, in an industrial society? They arise from the non-economic systems of society. Labor arises from the cultural system (e.g. families) of society. Natural resources arise from the science-technological system (e.g. environment) of society. Finances arise from the governmental regulated monetary system (central bank regulation) of society. *Polyani's point is that 'economic theory' by itself is insufficient theory about which to completely organize a stable industrial society.*

This may seem obvious today, but it was not in the history of ideas. Polyani wrote: "To allow the market mechanism to be the sole director of the fate of human beings and their natural environment . . . would result in the demolition of society. . . Robbed of the protective covering of cultural institutions, human beings would perish from the effects of social exposure; they would die as victims of acute social dislocation. . . A new class of employers was being created, but no corresponding class of employees could constitute itself. . . . Either machines had to be demolished . . or a regular labor market had to be created. Thus was mankind forced into the paths of a utopian experiment." (Polyani, 1944) "(Polyani, 1944) *The implementation of economic theory centered upon the concept of 'perfect market' provided a political ideology to lead industrializing nations toward an societal experiment in a utopia, Laissez faire economies.*

In the history of English industrialization, there had been a massive dislocation of the English population. Polyani wrote: "(There were major) vicissitudes to which the laboring people were exposed in England since the coming of the machine. First, (there were the vicissitudes of land enclosures) in the Speenhamland period, after 1795. Second (there were) the hardships caused by the Poor Law Reform, in the decade following 1834. Third (there was) the deleterious effects of a competitive labor market after 1834, until the 1870s when the recognition of the trade unions offered sufficient economy. Chronologically, Speenhamland antedated market economy; the decade of the poor Law Reform Act was a transition to that economy. The last period was that of market economy proper. . . . Much of the social upheaval and damage done to England's countryside sprang at first from the dislocating effects of trade directly upon the countryside itself." (Polyani, 1944)

The English industrial revolution historically proceeded in two phases. First there was rationalization of agriculture with new scientific practices, improving productivity and increasing income for English manors. To financially benefit from this, English lords enclosed previous common land, displacing peasants from traditional resources. This was called a 'revolution in agriculture, occurring in the early 1700s. Then it was followed the industrial revolution in the late 1700s.

Polyani wrote: "The Revolution in Agriculture antedated the Industrial Revolution. Both enclosures of the commons and consolidations into compact holdings, which accompanied the new great advances in agricultural methods, had a powerfully unsettling effect. The decrease of cottages, the absorption of cottage gardens and grounds, the confiscation of rights in the common deprived cottage industry (peasants) of its two mainstays: family earnings and agricultural background. As long as domestic industry was supplemented by the facilities and amenities of a garden plot, a scrap of land, or grazing rights, the dependence of the laborer on money earnings was not absolute; . .. The rationalization of agriculture inevitably uprooted the laborer and undermined his social security." (Polyani, 1944)

The previous 'Poor Law' of England had earlier helped societal stability, protecting paupers from starving (while economically not efficient for the owners of land and production). Polyani wrote: "In Canning's conviction, the Poor Law saved England from a revolution. He was primarily thinking of the 1790s and the French Wars. The new outburst of enclosures further depressed the standard of the poor in the countryside. . . . But for aid-in-wages the poor would have sunk below the starvation level in wide areas of rural England. Rick-burning was rampant. . . . Rioting was frequent. . . . (But) in 1834, industrial capitalism was ready to be started, and Poor Law Reform was ushered in. The Speehamland Law which had protected rural England, and thereby the laboring population in general, against the full force of the market mechanism was eating into the marrow of society. By the time of its repeal, huge masses of the laboring population resembled more the specters that might haunt a nightmare than human beings. . . The mechanism of the market was asserting itself and clamoring for its completion: human labor had to be made a commodity. . . Out of the horrors of Speenhamland, men rushed blindly for the shelter of a utopian market economy." (Polyani 1944)

This rushing toward the idea of a 'utopian' society of a market economy was seen as a way toward a prosperous future for a nation. But the new 'economic liberalism' would ignore important societal issues, such as was the distribution of wealth in utopia. In economic liberalism, there would be no principle of 'redistribution', as had been in tribal communities.

Both people and nature were to be seen only as productive inputs in economic liberalism. Polyani wrote: "Production in interaction of man and nature; if this process is to be organized through a self-regulating mechanism of barter and exchange, then man and nature must be brought into its orbit; they must be subject of supply and demand, that is be dealt with as commodities as goods produced for sale. . . Man under the name of labor, nature under the name of land, were made available for sale; the use of labor power could be universally bought and sold at a price called wages, and the use of land could be negotiated for a price called rent. . . While production could theoretically be organized in this way, the 'commodity fiction' (of labor and rent) disregarded the fact that leaving the fate of soil and people to the market should be tantamount to annihilating them." (Polyani, 1944) *People, as only a 'labor market', could be annihilated by 'unemployment'; and agricultural land, only as a 'rent, could be lost by erosion or paving in cities and suburbs (or by the rise of sea levels in global warming).*

The 'fictionalization' of inputs-as-kinds-of-produced-things is the 'utopian' idea – of an industrialized society as only an economy. Polyani wrote: "Our thesis is that the idea of a self-adjusting market implied a stark utopia. Such an institution could not exist for any length of time without annihilating the human and natural substance of society; it would have physically destroyed man and transformed his surroundings into a wilderness. Inevitably, society took measures to protect itself, but whatever measures it took impaired the self-regulation of the market, disorganized industrial life, and thus endangered society in yet another way." (Polyani, 1944)

A modern version of Polyani's concept of the 'impact-upon-nature-of-a- theoretical-economic-fiction' can be seen in this example of global warming – wherein conceptualizing nature's fossils (of coal and oil) only as economic commodities, over the last two centuries, has resulted in altering the natural environment through global warming. Economic thinking has proposed a 'carbon-tax' as a response; but a tax will not reduce pollution, only shift the costs of pollution. There is no 'economic' solution to global warming. There is only a technical one. Stop burning fossil fuels. To reduce global warming in the future, one technically needs not to burn fossil fuels, turning instead to nuclear fission to generate electrical power. This is why Polyani's criticism of 'economic fictions' can have real impact on society. *Materials and energy as natural resources are priced in markets but are not produced in markets. They are extracted from nature and, in the extraction, alter the natural environment.*

To offset the potential 'destructive aspects' of markets of labor, resources, and finance, Polyani argued that political institutions had to intervene. The political system needed to ameliorate the harsher aspects of these 'fictitious' markets. For example, child labor laws, and labor unions and government pension schemes protected the rights of workers in the labor market against exploitation, unfairly low wages and old age. Environmental laws protected the land and rivers from industrial pollution. Central banks and banking regulation protected the financial markets from inflation, bank runs, and fraud. Polyani wrote: "Factory legislation and social laws were required to protect industrial man from the implications of the commodity fiction in regard to labor power. Land laws ad agrarian tariffs were called into being by the necessity of protecting natural resources and the culture of the countryside against the implications of the commodity fiction in respect to them. Central banking and the management of the monetary system were needed to keep . . . enterprises safe from the harm involved in the commodity fiction as applied to money. Paradoxically enough, not human beings and natural resources only but also the organization of capitalistic production itself had to be sheltered from the devastating effects of a self-regulating market." (Polyani, 1944)

Polyani's critique was that traditional economic theory did not describe the whole needs of a society, only the economic needs. When economic theory completely dominates political theory in a nation, then the labor, land, and money in a nation can be abused. Polyani wrote: "(There were) two organizing principles in society. . . The one was the principle of economic liberalism, aiming at the establishment of a self-organizing market. . . The other was the principle of social protection, aiming at the conservation of man and nature as well as productive organizations . . . using protective legislation, restrictive associations, and other instruments of intervention as its methods. . . Two vital functions of society – the political and the economic – were being used and abused as weapons in the struggle for sectional interests. It was out of such a perilous deadlock, in the twentieth century, the fascist crisis sprang." (Polyani, 1944) *Two vital functions of society lie in its economic system and in its political system; and these must be coordinated in an efficient and just and patriotic partnership – for a society to thrive, avoiding the evils of dictatorship.*

LESSONS LEARNED

We have closely examined Polyani's thesis because it is the principle economic work which has seriously addressed the issue of itopian economics.

Proper economics and proper politics are both necessary for economic development.

Industry and government partnerships are essential to national economic development.

Market economics provides the means for industry to be efficient.

Regulatory politics provides the means for economics to be effective for societal prosperity.

Economic theory alone is insufficient to prescribe prosperous and stable capitalistic systems within a democracy. Complementary political theory is necessary to correct the inherent destabilizing 'fictions' in the markets of labor, resources, and finance. When political theory too much depends upon economic theory, then depressions occur in Laissez-faire economics democracies, such as the great depressions of the 1930s (which then encourage fascist movements). Economic theorists, such as John Maynard Keynes and Hyman Minsky, did address the instabilities inherent in financial markets. But the issues in the instabilities in the labor market and the resources market have not been fully addressed by all economic theorists. Political theorists too presently argue about the wealth inequity and the proper functioning of government and economy, particularly the influence of wealth on elections.

Next we will apply Polyani's analysis to the cases of North and South Korea; and we will see that:

> *South Korean society followed a historical path toward a balanced partnership between its political function and its Laissez-faire economics function. This government-industrial partnership led toward both economic wealth and democratic freedom.*
>
> *North Korean society followed a different historical path on a completely politically-controlled society. In a communist-military dictatorship, North Korean society became economically impoverished, militarily aggressive, and governmentally corrupt – with a citizenship either starved or imprisoned/exterminated.*

Chapter 12

DICTATORSHIP VERSUS DEMOCRACY – NORTH AND SOUTH KOREAS

INTRODUCTION

We have seen that both (1) a proper partnership between government and industry and (2) good leadership has been essential to the economic growth and success of a free South Korea. But the opposite has been the case of North Korea. Therein was no partnership between government and industry because North Korea was a communist government with no private industry. There was also bad leadership because this government was run by a dictator with a 'cult-of-personality'. As a result, North Korea has had economic collapse and government corruption – accompanied by the political oppression, widespread starvation, and imprisonment of its own people.

In 1956 at the Twentieth Party Congress of the Communist Party of the Soviet Union, Nikita Khrushchev had applied this term of 'cult-of-personality' to his Soviet Union predecessor, Joseph Stalin. Kruschchev described Stalin's terrible dictatorship, as focusing all power on his person, under which the Russia people were oppressed, starved, and imprisoned.

Societal structure and leadership are two essential features of modern industrialized society. The two Koreas show a stark contrast of structure and leadership between the same people. They had two different governments, dictatorship versus democracy, and two different economic systems, communism versus capitalism. How did North Korea get that way?

HISTORY: NORTH KOREA

Sixty years after the Korean war (which we earlier described), the conflict of ideology and government persisted between North and South Korea. There were periodic episodes of threats and hostilities. The latest (as this book was written) occurred in 2013. In March of that year, Rick Gladstone and David E. Sanger wrote: "Ignoring threats of retaliation, the United Nations Security Council ordered new economic sanctions against North Korea on Thursday for its third nuclear test last month, unanimously approving a resolution that the United States negotiated with China, the North's greatest protector. In an angry response, North Korea said

Friday that it was nullifying all agreements of nonaggression and denuclearization with South Korea and was cutting off the North-South hot line. . . . The new resolution instructs North Korea to cease all nuclear and missile testing and contains restrictions that will block financial transactions, impound cash, further empower countries to inspect suspicious North Korean cargo, and expand a blacklist of items that the North is prohibited from importing." (Gladstone and Sanger, 2013)

As we earlier reviewed, the civil war between North and South went back to Korea's division at the 38th parallel in 1945 by the Soviet Union and the United States, after the collapse of the Japanese Empire. Stalin had selected Kim Il-sung to head the communist North Korea territory above the parallel. Don Oberdorfer wrote: "In the North (Korea) Kim Il-sung systematically purged his political opponents creating a highly centralized system that afforded him unlimited power and generated a cult of personality." (Oberdorfer, 2001).

This is exactly what Joseph Stalin had done in Russia, centralizing power in the communist Soviet Union under his cult of personality. This is exactly what Mao Zedong had done in communist China, centralizing power in his cult-of-personality. This is exactly what Pol Pot did in Cambodia. And in each case, citizens were massacred, on vast scales (in the millions). This was the political reality of the communist regimes in the twentieth century – centralized power in a cults-of-personality – dictatorships not of the proletariat but of a person. Don Oberdorfer wrote: "As a leader, Kim was cordial to and comfortable with ordinary people. He emerged from a family of hard-working ordinary people and described himself in his memoirs as 'an ordinary man'. . . Yet at the same time, Kim came to live in luxury and exclusiveness beyond the dreams of kings. He inhabited at least five sumptuous palaces in North Korea and innumerable guest houses built for his comfort and amusement, all completely cut off from anyone except servants. . . . Like Stalin and Mao, whose cults of personality he emulated but far outdistanced, his automobile used special lanes, and other traffic was banished when he moved through the streets of his capital. . . . In deference to his health, a special institute was established in Pyongyang to concentrate on the aging process of this one man, with doctors and medial specialists monitoring his every move, and special fruits and vegetables produced solely for his consumption." (Oberdorfer, 2001)

But while the leader lived well, the ordinary North Koreans did not. A serious problem (besides injustice) with communist regimes is they are not economically efficient. For the North Korean economy to survive, Kim Il-sung needed subsidies. North Korea did not produce sufficient food or goods for its people. Kim went to the Soviet Union to obtain subsidies for his faltering country. In 1961, Kim signed a trade agreement with Nikita Khrushchev. Mao Zedong, the Chinese communist leader, did not like Khrushchev (because Khrushchev had denounced Stalin, a fellow leader whom Mao liked). Mao would not be outdone by Khrushchev and agreed to subsidize North Korea, just as magnanimously. Kim could live well.

So Kim Il-sung's North Korea had survived by subsidizes from both Russia and China; but times changed. In 1985, Mikhail Gorbachev became Chairman of the Communist Part of the Soviet Union. He tried to reform the Russian empire. One change was in relations with other communist countries. Seoul had sponsored the Olympics in 1988, and a month later Gorbachev began to do business with South Korea and began to reduce subsidies to North Korea.

The economic impact of change in Soviet Union's policies upon North Korea was huge. It had aided North Korea with $2 billion dollars annually, since 1984. These were in the form of loans, which were never repaid. The Soviet Union also provided North Korea with oil and gas and weapons. The trade was unbalanced, with the Russians shipping 1.9 billion dollars in goods to North Korea, while receiving goods in trade worth only 0.9 billion. North Korea had been a charity of the Soviet Union, from Stalin down to Gorbachev until 1989. But the charity ended. In December of 1988, the Soviet Union received a $300 million commercial loan from South Korea. The South Koreans began helping the Russians, rather than the Russians sustaining the North Koreans. The Soviet Union formally recognized the Republic of Korea and planned to continue to improve economic relations. In 1989, the Soviet Union collapsed. North Korea was on its own. North Korea had lost one major sponsor. Then China also stopped directly subsidizing North Korea, changing its trade to international prices.

Neither Russian nor China was subsidizing North Korea; yet instead of turning to economic reforms (as was occurring in China and Russia), Kim Il-sung decided to make nuclear weapons. Don Oberdorfer wrote: "In the early 1990s, North Korea's program to develop nuclear weapons concentrated the minds of many of the world's political and military leader . . . The more the outside world feared it, the more its nuclear program was a valuable asset to North Korea – which had few other resources of external worth, after the decline of its alliances with the Soviet Union and China. . . The first indigenous reactor at Yangban was a relatively modest plant . . . But in June 1988, a much larger reactor . . was photographed under construction at Yongbyon." (Oberdorfer 2001)

North Korea entered into negotiations with the United Nations for inspection of the nuclear facilities, claiming they were for electrical power and not nuclear bombs. Inspections were held but did not result in the amount of economic aid that North Korea anticipated. North Korea opened its nuclear activity to international inspections in 1992, but did not receive much international aid. In 1993, North Korea withdrew from participating the international Nuclear Non-Proliferation Treaty.

The economic crisis in North Korea continued, and in 1993, the North Korean Workers Party Central Committee publically admitted that targets in the North Korean economic plan were not being met. The North Korean government could not feed nor clothe its people, without foreign subsidies. Kim's policy response again was not reforming the economy but making nuclear weapons. In 1994, North Korea temporarily shut down its nuclear reactor to withdraw spent uranium fuel rods. From these rods, North Korean technicians removed enough enriched uranium material for an atomic bomb.

Then Kim Il-sung died. Don Oberdorfer wrote: "On the morning of July 6, 1994 . . . the 82-year-old 'Great Leader' was in good form, instructing senior officials on the economic goals for the year ahead. . . On July 7, en route to the mountains, Kim made one of his on-the-spot inspections of a collective farm. The temperature was nearly 100 degrees Fahrenheit. In the mountains, he personally inspected a guest villa, . . After these strenuous activities and his dinner, Kim complained of being tired. A short time later, he collapsed with a massive heart attack. . . Kim Il-sung was pronounced dead at two am. . . " (Oberdorfer, 2001)

While Kim had lived in wealth and luxury, his people had starved. In 1995, an official of the UN Food Program visited the countryside around Pyongyang and observed malnourished and hungry people everywhere. People were scavenging fields, looking for roots and wild plants to feed their starving children. In 1995, the North Korean government appealed for outside food aid.

It was not only in agriculture, but the whole of the communist economy of North Korea had collapsed, without its Soviet or Chinese subsidies. Don Oberdorfer wrote: "By the winter of 1996, most observers . . of North Korea had watched the progressive sinking of the economy . . . Deteriorating or flooded coal mines and reduced petroleum imports produced insufficient energy for industry, so many factories had closed or were operating at only a fraction of their previous output. . . Many office buildings and dwellings, even in the much-favored capital, were unheated during large portions of a very cold winter. Electrical blackouts were commonplace. . . North Koreans were consuming oak leaves, grasses, roots, tree bark. . with little nutritive value . . . Dormant factories were being dismantled and turned into scrap metal, which was then bartered across the Chinese border for the cheapest food available." (Oberdorfer, 2001)

Faced with a complete economic collapse, a North Korean government could have but did not chose reform. After the death of Kim Il-sung, the 'Great Leader', his the eldest son, Kim Jong-il, inherited the title of 'Dear Leader'. Kim Jong-il's policy was to continue confrontation with the international community, focused upon the issue of building a nuclear bomb. In September 1998, four years after the death of Kim Il-sung, the son, Kim Jong-il, was confirmed by the Supreme People's Assembly as the 'governmental leader of the country'.

Kim Jong-il secured his position through the support of the North Korean military. Don Oberdorfer wrote: "Although Kim Jong-il, unlike his father, had no military background, since his father's death he had spent a great deal of time establishing and improving close relations with the armed forces, the only people in the country capable of challenging him. Large numbers of officers had been promoted at his direction. . . General officers, of whom there were now many, were being driven around the capital by uniformed drivers in new Mercedes and BMW limousines. Despite the famine in the countryside, a special floor of the Koryo Hotel, the capital's best, had been set aside for the lavish wining and dining of senior military officers. Outside the capital, Russian-type dachas (recreational residences) were springing up for the use of military leaders." (Oberdorfer, 2001)

Kim Jong-il then turned to South Korea and the United States to subsidize North Korea, in return for allowing a nuclear inspection. In 1998, North Korea approved Hyundai's plan to build a resort in North Korea at Mount Kumgang. For this, Hyundai paid North Korean government $942 million dollars during the six years of its operation. The United States began to subsidize North Korea in return for nuclear inspection rights. North Korea shut down its nuclear reactors and received in return annual shipments of oil, worth nearly $100 million each year. International organizations began sending food aid to North Korea. Kim Jong Il's dictatorship survived. Kim Jong-il, had succeeded in obtaining subsidies from non-communist nations, for humanitarian and inspection reasons, while consolidating his political base with the North Korean army. But a belligerent stance had to be maintained, and in November 2010, North Korea fired artillery shells on Yeonpyeong Island. But on 17 December 2011, Kim Jong-il died of a heart attack.

One of his sons, Kim Jong-un, replaced him. International subsidies from communist allies had supported his grandfather, Kim Il-sung's regime. International subsidies from capitalist countries had supported his father, Kim Jong-il's regime. Why should not the policy of international subsidy continue? And North Korea's intimidation policy of threats-for food was tried again by the grandson. Kim Jung-un next took the opportunity of an annual joint military exercise between South Korea and U.S. forces for the occasion of renewing threats.

On March 1, 2013, the South Korean army and the U. S. army conducted its war exercise. On March 7, 2013, the North Korean government announced an intention to eventually launch a preemptive nuclear strike against the United States. It withdrew from pacts with South Korea and closed the border crossing at the 38th parallel. It ended the 1953 armistice agreement with the United Nations. At the end of March 2013, North Korea declared it was in a state of war with South Korea. In May, North Korea fired four short-range missiles into the sea off Korea.

Once again, the time was tense between the North and the South Korea. In March 26, Choe Sang Hun wrote: "North Korea's military said it put all its missile and artillery units on 'the highest alert' on Tuesday, ordering them to be ready to hit South Korea, as well as the United States and its military installations in Hawaii and Guam. The threat from the North's Korean People's Army Supreme Command came only hours after President Park Geun-hye of South Korea warned that the North Korean leadership could ensure its survival only when it abandons its nuclear weapons, long-range missiles, provocations and threats. North Korea has since issued a torrent of threats to turn Washington and Seoul into a 'sea of fire.' Its leader, Kim Jong-un, who has inherited the 'military first' policy of his late father, Kim Jong-il, has made a round of visits to military units in the last week. He inspected live-fire artillery and amphibious landing exercises, ordering his soldiers to send the enemies 'to the bottom of the sea as they run wild like wolves threatened with fire,' according to North Korean media." (Choe, 2013a)

Things were tense all through March 2013. Choe wrote: "Although North Korea issued strident threats and stirred up fears of American invasion during previous joint American-South Korean military drills, Mr. Kim has been far more aggressive in issuing such threats personally than was his late father, Kim Jong-il. Unlike his father, who had expanded his power base from his youth, Mr. Kim was catapulted into top leadership after his father's sudden death in 2011 and must build his credentials as head of his 'military-first' government, South Korean analysts and officials said. Hours after Mr. Kim's call to arms, thousands of North Koreans turned out for a 90-minute mass rally at the main square in Pyongyang, chanting 'Death to the U.S. imperialists' and 'Sweep away the U.S. aggressors,' according to The Associated Press, which has a bureau in Pyongyang. Soldiers and students marched through downtown Pyongyang. On Thursday, the American military carried out a rare long-range practice bombing run over the Korean Peninsula, sending two nuclear-capable B-2 stealth bombers on a practice sortie over South Korea, underscoring Washington's commitment to defend its ally amid rising tensions with North Korea." Choe, 2013a)

The tension triggered a policy shift. The United States and South Korea moved toward China, for the three together to rein in North Korea's threat. Mark Landler wrote: "The Obama administration, detecting what it sees as a shift in decades of Chinese support for North Korea, is pressuring China's new president, Xi Jinping, to crack down on the regime in Pyongyang or face a heightened American military presence in its region. In a flurry of exchanges that included a recent phone call from President Obama to Mr. Xi, administration officials said, they have briefed the Chinese in detail about American plans to upgrade missile defenses and other steps to deter the increasingly belligerent threats made by North Korea's young leader, Kim Jong-un. China, which has been deeply suspicious of the American desire to reassert itself in Asia, has not protested publicly or privately as the United States has deployed ships and warplanes to the Korean Peninsula. That silence, American officials say, attests to both Beijing's mounting frustration with the North and the recognition that its reflexive support for Pyongyang could strain its ties with Washington." (Landler, 2013)

On September 4, 2013, The Korea Times Herald reported on the South Korean government's assessment of North Korea's nuclear capability: "North Korea has made considerable progress in developing a fairly robust nuclear program in the past three years and is capable of making atomic weapons at any time, Seoul's defense ministry said Tuesday. 'North Korea's nuclear program remained at a developmental and experimental stage in 2010, but it has grown into a real threat in 2013, and it is able to make a nuclear device at any time." (Yonhap News 2013).

On the topic of political imprisonment of North Korean people, North Korean prison camps, a Washington Post reporter, Blaine Harden wrote: "North Korea officially says the camps do not exist. It restricts movements of the few foreigners it allows into the country and severely punishes those who sneak in. U.S reporters Laura Ling and Euna Lee were sentenced last month to 12 years of hard labor, after being convicted in a closed trial on charges of entering the country illegally." (Harden, 2009)

Yet both defectors and satellites provided evidence of the existence of the prison camps. Blaine Harden wrote: "Images and accounts of the North Korean gulag become sharper, more harrowing and more accessible with each passing year. A distillation of testimony from survivors and former guards, newly published by the Korean Bar Association, details the daily lives of 200,000 political prisoners estimated to be in the camps. Eating a diet of mostly corn and salt, they lose their teeth, their gums turn black, their bones weaken and, as they age, they hunch over at the waist. Most work 12 to 15-hour days until they die of malnutrition-related illnesses, usually around the age of 50. Allowed just one set of clothes, they live and die in rags – without soap, socks, underclothes, or sanitary napkins. The camps have never been visited by outsiders, so these accounts cannot be independently verified. But high-resolution satellite photographs, now accessible to anyone with an Internet connection, reveal vast labor camps in the mountains of North Korea. The photographs corroborate survivors' stories, showing entrances to mines where former prisoners said they worked as slaves, in-camp detention centers where former guards said uncooperative prisoners were tortured to death and parade grounds where former prisoners said they were forced to watch executions. Guard towers and electrified fences surround the camps, photographs show." (Harden, 2009)

About the conditions of the political prisons in North Korea, in 2011, Amnesty International made an appeal: "Amnesty International is releasing new footage and testimony to highlight the horrific conditions of political prison camps in North Korea at a time when there is fear of a potential new crackdown in preparation for a possible succession in leadership. The basis of the briefing comes from testimony from 15 former inmates and prison guards – a majority of whom have decided to remain anonymous for their own security. Amnesty International is calling on the North Korean leader to: (1) Acknowledge the existence of political prison camps in North Korea; (2) Immediately close all political prison camps in North Korea; (3) Immediately and unconditionally release all prisoners of conscience, including relatives held on the basis of 'guilt-by-association'; (4) Release all other inmates unless they are charged with an internationally recognizable offence, remanded by an independent court and given a fair trial." (Amnesty International, 2011)

Amnesty International estimated about 200,000 prisoners were held in six camps, including Camp 14 in Kaechon, Camp 15 in Yodok and Camp 18 in Bukchang. People were arrested and deported to the camps without trial, along with their family. It estimated that over 10,000 people died each year. (Amnesty International. 2011)

Blaine Harden wrote: "They have existed for half a century, twelve times as long as the Nazi concentration camps and twice as long as the Soviet Gulag. Although precise numbers are impossible to obtain, Western governments and human rights groups estimate that hundreds of thousands of people have died in the North Korean camps." (Harden, 2009)

SOCIETAL DYNAMICS ANALYSIS OF THE TWO KOREAS

From 1946 to 2013, the divided Korea had evolved into two societies: South Korea as democratic and prosperous and free, and North Korea as dictatorial and impoverished and imprisoned. What can their different societal experiences tell us about the modern global civilization? We next analyze the structure and leadership of the two Koreas in the framework of societal dynamics, Figure 12.1.

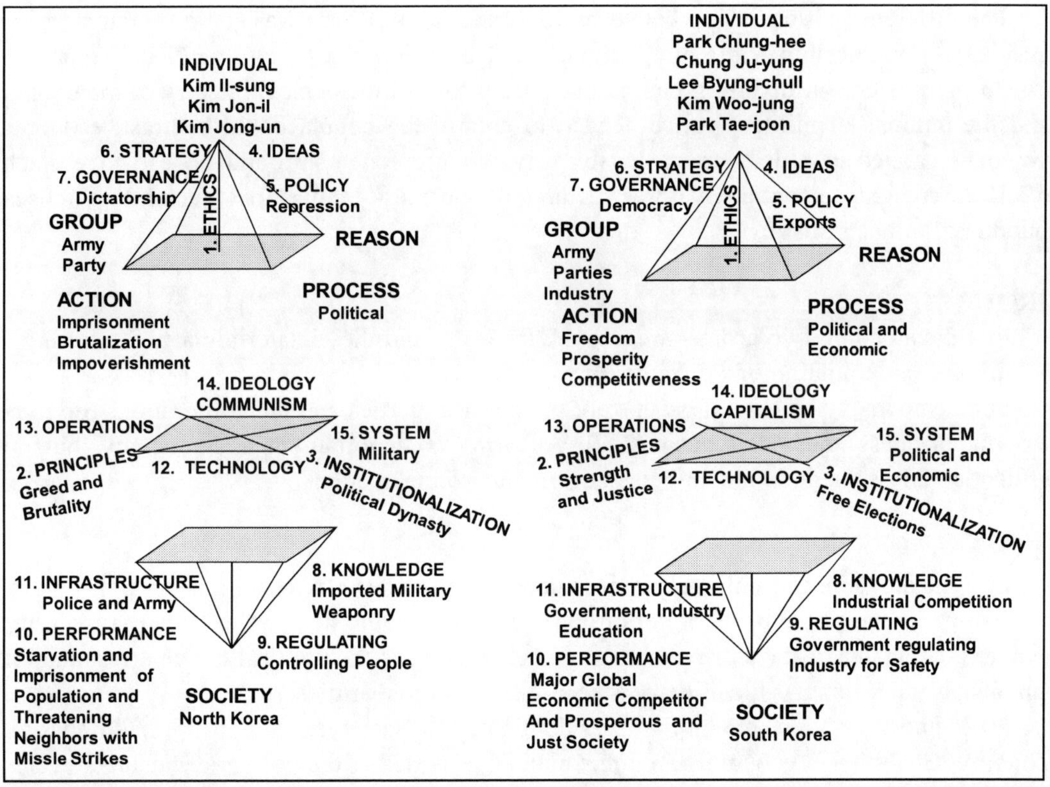

Figure 12.1. Two Koreas.

SOCIETY –

From 1946 to 2013, North Korea developed as a subsidized communist dictatorship, becoming one of the poorest and most brutal countries in the world. Population starved and prison camps were responsible for imprisonment and deaths of millions of North Koreans. All the while, government and military officials lived in luxury, being driven around in foreign cars.

From 1963 to 2013, South Korea developed from one of the poorest countries in the world to one of the economically most prosperous and competitive countries. Hyundai Motor became the fifth largest automobile company in the world, exporting around the world.

INDIVIDUALS –

The key individuals in the development of North Korea were three communist dictators: Kim Il-sung, followed by his son, Kim Jong-il, and his son, Kim Jong-un. They were dynastic rulers of the Democratic People's Republic of Korea (DPRK).

The key individuals in the development of South Korea were in government and in industry. In the government of the Republic of Korea (ROK), there had been President Park Chung-hee, who was followed by eight other presidents (with the last seven elected democratically). Other key individuals were founders of chaebols and industrial leaders, including: Chung Ju-yung (Hyundai), Lee Byung-chull (Samsung), Kim Woo-jung (Daewoo), and Park Tae-joon (Posco).

The difference in leadership between North and South Korea was stark: dictators versus presidents-in-partnership-with-entrepreneurs. Dictatorship does not facilitate economic growth but prevents it, diverting national resources toward the personal luxury of the dictator (and the minions of military/police needed to control the populace). In contrast, economic growth is created in a democracy, led by private entrepreneurship and honest government officials, creating a national infrastructure of business, education, technology, fiscal soundness, and patriotism.

GROUP

In a dictatorship, two groups are central, the army and the dictatorial party. The army is the ultimate underpinning of a dictatorship.

In a democracy, the central groups are the political parties and the corporations (dividing into big business and small businesses). An army is important in a democracy, but not influential unless the government breaks down and rioting occurs.

PROCESS

In a dictatorship, the political process dominates the society, centering upon a 'cult-of-leadership' and bolstered by a core of military leaders and a political police (internal security) loyal to the dictator. A dictator remains in power, as long as the political police arrest and kill opponents; and as long as the military is secure from foreign armies.

In a democracy, the economic process dominates the society; and industry determines the economy. Government economic stimulus may be provided at the bottom of business cycles. Unregulated financial systems periodically create financial bubbles which can create bank runs, which create economic depressions. Government processes should focus on societal infrastructure, such as education, health, pensions, energy, justice, regulation for safety and honesty, – all under a rule of law.

REASON

The reasoning of the North Korean regime was that government should run the economy (communism), and that the primary beneficiaries of the economy should be the party and military (dictatorship).

The reasoning of the South Korean regime was that government should work in partnership with private industry which runs the economy, and that the military should serve to protect the country.

ACTION

From 1946 to 2013, the actions of the North Korean regime was to imprison , brutalize and impoverish its people, while begging foreign governments to subsidize its economy

The actions, from 1946 to 2013, of the South Korean regime was to provide freedom and prosperity for its people, while building an internationally competitive society.

LESSONS LEARNED

From this comparative analysis of North and South Korea, one sees evidence about two basic societal principles: dictatorship versus democracy and communism versus capitalism. For the government of a people in a nation, dictatorship is bad and democracy is good. For the economic system in a nation, communism is ineffectual and capitalism is efficient.

A government and economy of dictatorship and communism has historically been bad not only for its own people but for its neighbors, as evidenced in the Soviet Union under Stalin and Communist China under Mao and Cambodian government under Pol Pot (and other examples). Communist and fascist dictatorial regimes of the twentieth century all had brutalized, enslaved their people, and killed large numbers of populations – all without creating economically viable economies. Communism didn't work economically; and communist and fascist dictatorships never provided justice and security.

Democratic governments with capitalistic economies have provided freedom and relative wealth for their peoples.

But there still remain serious problems in democracy and capitalism: (1) of how to ensure an honest government in a democracy and (2) of how to provide a stable economy in a capitalist society, while (3) preserving the environments of the world.

Chapter 13

SELF-ORGANIZING SOCIETIES – KOREA

INTRODUCTION

Economic development is not simply a matter of individual leadership; nor is economic development simply a matter of social structure (a Laissez-faire free-market). Economic development requires both, leadership and structure. However, good leadership and proper societal structure together are not simple processes, nor certain. Bad leadership can corrupt societal structure. Corrupt societal structure can promote bad leadership.

We saw this in North Korea. Bad leadership, in the dictatorship of Kim Il-sung, created a communist society that was inefficient and ineffectual, neither feeding nor securing its people. Moreover, the communist society promoted corrupt leadership as military and government officials continued the dictatorship through Kim's son and grandson. But even where there has been good leadership and proper structure, as in South Korea, there was a delicate balance between leadership and structure to ensure that: (1) a capitalistic society creates prosperity that is widespread over both capitalists and labor and (2) a democratic society elects governments which rule for the public good rather than private interests. Good leadership in a good societal structure is a balance, but not an easy one.

Accordingly, we have analyzed the economic development of South Korea within this basic sociological dichotomy of *Individual and Society*. Society is composed of individuals, and individuals act within the context of society. Economics consists of agents acting within the structure of a market. The economic agents are individuals, and markets provide the social structure in which their actions occur. We saw this interaction of individual and society in the modern history of South Korea – that much of its national success depended upon key individuals displaying outstanding business and government leadership to build an industrialized export-oriented economy – within a capitalistic democratic society – leadership and structure.

HISTORY: LATE JOSEON SOCIETY IN KOREA

Earlier we reviewed how Karl Polyani described the societal disruption in England that accompanied its historic movement toward industrialization. Societal dislocation has accompanied the beginning of industrialization in all cultures. And this dislocation has not been a simple thing of 'structure' in society but also has involved the kinds of leadership in society at the time. To see how corrupt individuals can facilitate an unjust society during economic dislocations, we next review the long ago time in Korea just before the Donghak Rebellion in 1894.

We recall that this revolt had caught up Park Chung-hee's father and resulted in his being disowned. Park then grew up in poverty, disgusted at a corrupt and ineffectual society. When Park became president, he was determined to develop South Korea into a strong, prosperous and free nation, with justice for the Korean people. Shin Soon Chul and Lee Jin Young described the historical societal conditions in Korea in the 1800s: "Entering into the 19th century, the Joseon dynasty underwent great economic changes resulting from the successful introduction of a new method of agriculture in which rice seedlings were transplanted to the rice paddy, a method which wholly replaced the traditional direct planting . . . Universal adoption of this method . . . brought many advantages to farmers, including a more than 50% decrease in required labor and a two-fold increase in crop yields . . . Transplanting allowed farmers to double their yearly production by alternating crops of rice and barley in a single paddy. . . In addition, remarkable expansions and improvements in water storage and irrigation facilities, new developments in fertilizing methods and farming equipment, the introduction of crops like the potato, sweet potato, and chili pepper , and expansion of crops like ginseng and tobacco also contributed to growth in agricultural productivity." (Shin and Lee, 2008)

This was remarkable improvement of agriculture, as knowledge from the foreign sources was being imported into the previously closed Korean society. Shin and Lee wrote: "The boom in agricultural production led to changes in agricultural management. A smaller labor force and increased crop yields allowed the landlords and the peasants to expand their scope of operations. To increase land holdings, they reclaimed unfarmed land, purchased new land, or leased Iand from others for cultivations. As productivity grew, a commercial monetary economy rapidly developed, and the agricultural economy became linked to markets. Grains, textiles, vegetables, medical herbs, tobacco, etc. were now grown not only for consumption but also for sale in the market." (Shin and Lee, 2008)

Classical economic theory was partly working in Joseon Korea, as foreign knowledge was being imported. As supply expanded, demand expanded and Korean markets developed, along with a growing financial market. Agriculture technology was stimulating productivity which stimulated economic growth. And then (as Polyani argued), a labor market developed. Shin and Lee wrote: "The peasants participating in the new commercialized form of agricultural management were either independent farmers who owned their farmland, or tenant farmers who leased land from others. Some of these farmers amassed wealth, and became well-to-do farmers or peasant landlords. But as the prosperity of the new expanded system put the majority of property and wealth into the hands of a minority of landlords and merchants, many peasants ended up losing their land. In short, in the process of changes in the agricultural economy during late Joseon, the peasant class became increasingly polarized between a minority of wealthy farmers and landowners and a majority of impoverished farmers." (Shin and Lee, 2008)

Traditional economic theory would have argued that any market (including a food market) should have become perfect – consequently and automatically bringing prosperity to Korean farmers. We recall this was Karl Polyani's criticism of 'Laissez fair' economic theory as a political ideology – that democracy need not automatically follow upon a free market.

Historically, the Korean agricultural market did not become 'perfect'; and 'democracy' did not occur in the late Joseon dynasty. Both leadership and structure prevented it. Over the course of the nineteenth century, peasants became poorer, government more corrupt, and land owners more wealthy (but only temporarily). In 1894, as the century drew to its end in Korea, the Donghank Rebellion and the beginning of Japanese occupation occurred.

Where in classic economic theory is there any thesis that wealth will be fairly distributed in an industrial economy? Traditional economic theory is not much concerned about the distribution of wealth, only about the efficiency of pricing. (Many economists don't discuss wealth distribution, but some do.) In particular, Laissez-faire economic ideology has paid little attention to uneven wealth distribution. This was one of the weak points of classical economic theory – upon which jumped Karl Marx and his communist followers, assuming capitalism would collapse because workers would be starved to death. Of course, historically this never happened. Workers formed unions and negotiated on wages with management (as Polyani pointed out).

But back in late Joseon Korea, industrialization had not yet started, and poor farmers had no effective organization (like an industrial union) to protect their economic interests – hence the peasant rebellions. Shin and Lee wrote: "Polarization increased as time passed. The rich farmers hired labor for production and sold produce at the market for currency, which they used to further expand their landholdings. Thus by the latter half of the 19th century, developments in agricultural production resulted in a polarization within the farming class, a situation which rocked the foundations of the entire society." (Shin and Lee, 2008)

We also recall that this was Karl Polyani's point about the economic 'fictions' of labor and land, priced as a kind of 'market'. Neither the labor nor the land in Korea was produced. Korean farm children were born, a few into rich families but a majority into poor families. Labor was not merely a 'market' with a daily price for labor; but it was also a polarizer of society, becoming a tension in the social bond between Korean people. We recall Polyani wrote: "In addition to the principle of 'reciprocity' in non-industrial society, a second principle of 'redistribution' of goods does smooth some of the unevenness of wealth in a tribe." (Polyani, 1944) But in the beginning of any economic 'industrialization', tribal principles of' reciprocity' and' redistribution' of societal wealth are historically lost. This happened in England, in Europe, in America, in Japan. And it was happening in nineteenth century Korea, when industrializing societies were beginning to engage with Korean society. Korean agricultural technology benefited the Korean peasant community, but the community was divided into a small number of the wealthy and a large number of the impoverished.

According to Polyani's thesis, government should then have stepped into balance the society in social justice. We recall Polyani wrote: "Our thesis is that the idea of a self-adjusting market implied a stark utopia. Such an institution could not exist for any length of time without annihilating the human and natural substance of society; it would have physically destroyed man and transformed his surroundings into a wilderness." (Polyani, 1944).

So the Korean government needed to step into the changes going on in Joseon Korea to avoid annihilation of the human and natural substance of its society. But instead, the Korean government became even more of a problem to Korea, accelerating the destruction of the 'human substance of society', the peasants. Shin and Lee wrote: "During this period, taxes were levied in a concentrated manner on the commoners, the independent and tenant farmers. Land taxes were not levied on the landowners, but on the people who cultivated the land. To make matters worse, from the mid 18th century, Joseon's tax collection system was organized under a three-tier administration, with separate divisions to oversee land taxes (jeonjeong), military duty (gunjeong), and repayment of borrowed government grain (hwangok). But the system harbored problems and contradictions. The basic contradiction lay not only in 'how much' taxes the government exploited, but in the 'who' they exploited." (Shin and Lee, 2008) Joseon officials taxed the poor.

In taxation, it is important not only how much is paid but who pays it and whether the tax collectors are honest. Shin and Lee wrote: ". . . the work of collecting taxes was left completely in the hand of the local Governor and his clerk who in conspiracy with Central Government officials indulged in freely exploiting the peasants. Local officials might put abandoned fields on the roster and levy land taxes on them . . . Taxes were not supposed to be levied for 3 to 5 years on land newly brought under cultivation, but corrupt officials registered it as taxable land anyway. They applied miscellaneous taxes for anticipated rice shortages, anticipated rice spillages . . . resulting in the peasants having to pay many times more than what was laid down by the rules. . . . . For example in 1893 in the Gochang region of Jeolla Province, 30% to 40% of tenant farmer households had barely enough grain to last for 4 months, after a year of farming and after all their taxes were paid; and 35% of the land-owning peasant had enough gain of last for only 7 months. Because of the brutal tax policy, 70% of the population was unable to stock enough food for a year." (Shin and Lee, 2008)

A free-market economy requires a government context of incorruptible officials, who act in the public good and not in their private good as dishonest officials. A central Korean government in the 1800s was not acting in the public good of the Korean people but in their private good of seizing power and wealth. Shin and Lee wrote: "In the beginning of the 19th century during the reign of King Sunjo, a small fraction of people related to the Royal family attempted to seize power by unlawful means. Such illegal grabs for power were not limited to a single instance, but continued . . . Decades later, when King Gjong succeeded to the throne, the young king's father, Yi Haeung, himself assumed direction of the Government and set in motion a resolute program of reforms aimed at creating a strong monarchy. He sought to destroy the pattern of rule by a few powerful clans . . . unjustly monopolizing Royal sovereignty. But after 10 years, Yi retired and the Queen's relatives . . . recommenced 'In-Law Government'. The influences of the Min family were attended by many abuses that led to crisis in the Joseon government. . . opening up the country's ports to the Japanese and . . . and making China its suzerain during the military mutiny of 1882 and Coup d'état of 1884." (Shin and Lee, 2008)

Thus Shin and Lee had described (1) the beginnings of a economic free-market economy in agriculture, (2) accompanying polarization of the peasantry into wealthy land-owners and impoverished workers, (3) corrupt local government officials exploiting peasants with excessive taxation, (4) incompetent central government of clans struggling for control of royal power, and (5) foreign powers' domination over Korean government. The societal context of the development of a free-market industrialized society in Korea failed, partly due to a government, which was corrupt, incompetent and powerless.

When industrialization begins in a nation, a strong, competent, and honest government must develop proper societal infrastructure and administer justice. Otherwise a society is open to rebellion and/or foreign domination. Shin and Lee wrote: "In 1876, Joseon succumbed to threats by the Japanese military and opened up the country's ports with absolutely no advance preparations. After the opening of the ports, Joseon began trade with Japan. Joseon's exports included agricultural and fishery products like rice, beans, and cowhides, and precious metals like gold. . . . The bulk of imports to Joseon were Japanese cotton and daily necessities. This foreign trade structure spelled doom for the foundations of Joseon's agriculture and handicraft industries. . . . In the face of massive rice exports, Joseon, which had been suffering from rice shortages to being with, now experienced critical rice shortages that led to price increases. . . It was the same case with the handicraft industry. It too was ruined as cheaper cotton products from abroad wiped out the native home-based cotton market. . ." (Shin and Lee, 2008)

Also we recall that Karl Polyani wrote about the importance of the political context of economic systems: "To allow the market mechanism to be the sole director of the fate of human beings and their natural environment . . . would result in the demolition of society. . . Robbed of the protective covering of cultural institutions, human beings would perish from the effects of social exposure; they would die as victims of acute social dislocation. . ." (Polyani, 1944) In case of Joseon Korea, Shin and Lee wrote: "Peasants unable to tolerate the poverty and exploitation could show either passive resistance by leaving their villages and wandering, begging for food, or active resistance, either through legally submitting written complaints about their living conditions to the officials or by mobilizing their physical bodies to rise up in revolt. Armed revolt was the most popular way for the farmers to express resistance. Revolts . . . were especially concentrated in the latter half of the 19th century . . . Indeed, matters only worsened leading up to the year of 1884." (Shin and Lee, 2008) This was the background to the Donghak Peasant Revolution in 1884.

JOSEON SOCIAL STRUCTURE AND INDIVIDUAL ETHICS

What we see in this historical example is that how a societal structure actually performs depends both upon the socio-technical systems of the structure and upon the ethics of individual leadership in that structure, Figure 13.1.

The economic *System* of Joseon society was beginning as a free-market economy, based upon new *Technology* of agriculture (in the transplanting of rice seedlings). This increased the demand for land upon which to produce rice, pricing land beyond the reach of ordinary peasants and for only the wealthy peasants and yangban (landed aristocracy). This increased the polarization in the *Groups* between the wealthy and the poor. This polarization fomented *Actions* of the recurrent peasant rebellions. These rebellions were encouraged by the bad *Ethics* of corrupt governors, extorting excessive taxes on the peasant producers of agricultural products. Due to bad *Governance,* the central government did not properly *Regulate* officials and prevent bad behavior of local governors; and the *Governance* of the royal family was internally preoccupied by clans struggling over the royal power. The result of the social dislocation and corruption was a societal *Performance* that made Korea vulnerable to foreign power domination.

When government corruption and bad governance do not address issues of societal dislocation (resulting from economic development of free markets from new technology), then the whole society can be weakened, resulting in internal rebellions and/or foreign domination.

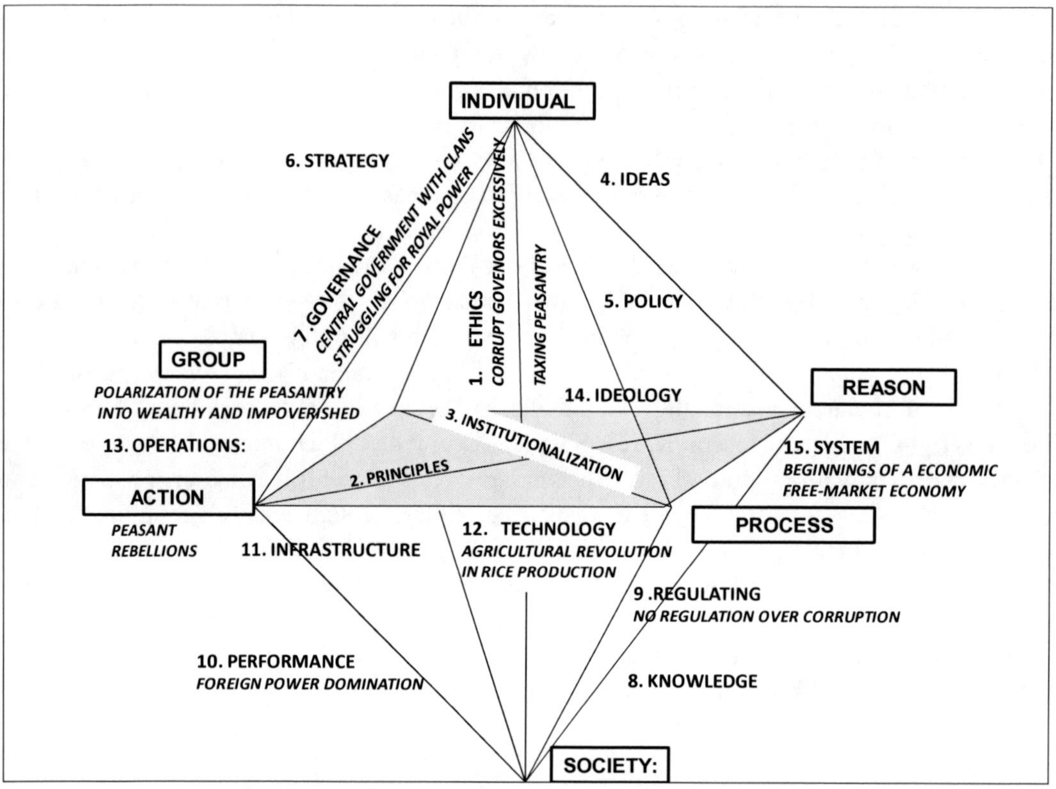

Figure 13.1. Changes in Nineteenth Century Joseon Society.

SOCIAL STRUCTURE AND LABOR MARKETS

Compared to the availability of land for peasants in a feudal society, like the Joseon society, in a modern industrialized society, the equivalent basis of societal stability is 'employment' – the pricing of labor and the availability of jobs in the industrialized economy. Thus employment is a key factor in the prosperity and stability of an industrial society. The measure of 'unemployment' in a nation is a critical measure of economic health.

For example in 2011, Claire Melamed, Renate Hartwig and Ursula Grant wrote: "Jobs, and in particular the quantity and quality of new jobs created by economic growth, are shaping up to be one of the crucial development issues. . . . The International Labor Organization (ILO) estimates that 440 million new jobs need to be created in the next ten years to keep up with population growth and demographic changes. It's not just 'creating jobs' that is the challenge. Nearly half of all workers worldwide still live below the $2 a day poverty line. Jobs must also pay enough to end poverty. The key determinants of the relationship between growth, poverty reduction and inequality are whether economic growth generates new jobs, the quality of these jobs, whether poor people are able to take up new opportunities, and whether jobs are stable enough to last in the face of economic shocks. But we know surprisingly little about the empirical relationships between growth episodes and different types of employment, and even less about what policies and programs might enhance positive linkages between the two. This is an urgent policy question, since economic growth has a very mixed record in job creation. (Melamed et al, 2011)

Melamed, Hartwig and Ursula described in 2011: "Low-skill manufactures are losing their place as the drivers of job creation, while services are becoming more important. As the world becomes less secure, the shift from agriculture to manufacturing is no longer a one-way street. The trajectory from growth to employment to poverty reduction is not automatic. Distinctions between formal and informal sector firms are breaking down. Employment patterns are being shaped by two major inequalities: between young and old, and between men and women." (Melamed et al, 2011)

This is now a critical labor issue throughout this global industrial civilization. Manufacturing no longer provides jobs for low-skill workers. Over the course of the twentieth century and continuing into the twenty-first century, automation technology in manufacturing has been taking out low-skilled jobs.

> For example in 2013, Stephanie Clifford wrote about textile manufacturing in the United States: "The old textile mills here are mostly gone now. Gaffney Manufacturing, National Textiles, Cherokee (former clangorous, dusty, productive engines of the Carolinas fabric trade) fell one by one to the forces of globalization. Historically, North Carolina benefited after manufacturing migrated first from the cotton mills of England to the mill towns of New England and then to North Carolina) - where labor was even cheaper. But in the 1990s, the textile industry mostly left the United States. It headed to China, India, Mexico — wherever people would spool, spin and sew for a few dollars or less a day. Which is why what is happening now at the old Wellstone Spinning Plant is so remarkable. . . . Parkdale Mills, the country's largest buyer of raw cotton, reopened it in 2010." (Clifford, 2013)

Labor was a significant cost in textile manufacturing, when the technology was only semi-automated; and then textile manufacturing moved from higher-labor-cost areas to cheaper, from England to New England to North Carolina to India. But as automation continued, manufacturing could move back to a higher-wage location, because labor costs are not a significant factor in a highly-automated industry.

> Stephanie Clifford wrote: "Bayard Winthrop, the founder of the sweatshirt and clothing company American Giant, was at the mill one morning earlier this year to meet with his Parkdale sales representative. Just last year, Mr. Winthrop was buying fabric from a factory in India. Now, he says, it is cheaper to shop in the United States. Mr. Winthrop uses Parkdale yarn from one of its 25 American factories, and has that yarn spun into fabric about four miles from Parkdale's Gaffney plant, at Carolina Cotton Works. Mr. Winthrop says American manufacturing has several advantages over outsourcing. Transportation costs are a fraction of what they are overseas. Turnaround time is quicker. Most striking, labor costs (the reason all these companies fled in the first place) aren't that much higher than overseas because the factories that survived the outsourcing wave have turned to automation and are employing far fewer workers. . . . (Also) monitoring worker safety in places like Bangladesh (where hundreds of textile workers have died in recent years in fires and other disasters) has become a huge challenge. . . ." (Clifford, 2013)

In the twenty-first century, manufacturing was not providing sufficient numbers of jobs in an industrialized nation for full employment. Some scholars were already envisioning that the industrialization patterns of the past may need to be changed. For example, Yuting Qiu and Fan Yiyang wrote: "The prosperity of emerging markets is challenging the traditional global innovation strategy. The unique characteristics (i.e., market fragmentation, demand heterogeneity, shortage of resources and inadequate infrastructure) of emerging markets impel their emergence as a source of innovation and disruptive force. Only (different) innovation . . . can match these characteristics and contribute to the rest of world. . . . The essence of innovation should be solving everyday problems, eradicating poverty and hunger, and helping the people attain an inclusive-and-sustainable future." (Yuting and Fan, 2013)

What is the proper concept for 'economic efficiency'? Is it merely the reduction of the cost of labor in production? Perhaps economic efficiency should be thought about as a more complex idea in a modern global civilization. What if the idea includes that of a reduction of fossil fuels in production, for example? Or what if it includes the idea of conservation of agricultural land, for another example? Or what about the concept of 'economic efficiency' also including the idea of adjusting employment opportunities to family structures? Perhaps, traditional economics has thought too simply about what is really 'efficient' economically for a prosperous and secure society.

As new economists, Yuting and Fan were urging this kind of re-thinking the concept of 'economic productivity', beyond that of merely a focus upon capital and labor. They wrote: "Three major revolutions (agricultural, industrial and information revolutions) in past centuries determined the course of history. Every revolution exhibits its own distinctive traits. For example, the Agricultural Revolution made it possible to meet the basic food needs of large population.

"The Industrial Revolution affected the production process disruptively. The Information Revolution changed the way people think, understand and connect with the world. All in all, they as a whole created the modern society we live in today. . . . In many populous countries, like China and India, the path to an inclusive-and-sustainable future is a long journey, full of obstacles. They have to battle growing pains in the process of economic development, such as great inequality and pollution of environment. But the primary task is still eradicating poverty and hunger.

"Two pragmatic approaches can achieve this goal. First, fully deploy social resources to target specific areas such as consumer literacy, micro-financing, and the supply of basic necessities. Second, encourage public–private partnership to focus on the societal needs of un-served markets on a sustainable long-term basis." (Yuting and Fan, 2013)

Yuting and Fan were urging looking at an industrial society, not merely as a production system but as a production and consumption system together. In this way, the economic concept of a 'free-market' is not merely an optimizer of commodity 'prices', but also an optimizer of 'employment' and 'basic societal necessities'. This kind of economic idea provides a larger 'context' to the idea of a free-market – not only in meeting supply with demand but also in enabling a consumer society.

Corrupt leadership distorts and limits economic growth; and unjust societal structures impede and disrupt economic growth. Both good leadership and proper societal structure are necessary for effective economic growth. But how are the conditions for the 'good' in leadership institutionalized? And how is corruption in a social structure limited, identified and corrected?

HISTORY: HYUNDAI LABOR UNREST

We have looked at the peasant's rebellion in 1894. We saw that this occurred as a result of the societal polarization between the Korean wealthy and the Korean poor (with wealth as land-ownership). This contributed to the weakening of the Korean society and to the subsequent military occupation by the Japanese army.

Now even in a modern society, there are such conflicts, between the wealthy and the poor, between the powerful and the powerless. In 1989 in modern South Korea, this conflict emerged between management and labor in the chaebol corporations. It was in a different time and in different conditions; but it was a similar conflict of societal polarization, about in a society who gets how much.

For example, there was a series of labor strikes in the Hyundai Group of companies. Donald Kirk wrote: "Several weeks after Park's assassination in October 1979, a worker at Hyundai Heavy Industry (HHI) was arrested after loaning a friend a book about the rise of the American labor movement. The court found him not guilty, but HHI forced him to resign. . . . The realization that Hyundai was earning tremendous profits in an era of unparalleled prosperity for Korea contributed to the (labor) movement. An activist Kwon wrote: 'The annual growth of Hyundai was 30% but our raise was always around 3% or 4%. . . . (We) activists decided the time had come to form a union and make formal demands . . . Since Chung Ju Yung's policy was never to give permission for a labor union, we thought of the strike and demonstration as our only solution.'" (Kirk, 1994)

The recognition of unions did not occur easily at Hyundai. About ten years later in 1988, strikes occurred at Hyundai. Donald Kirk wrote: "The time was 10:30 pm on a balmy spring evening, May 6, 1988. The place was Seoul's Gangnam district, a relatively new area of fashionable apartment blocks and office buildings, nightclubs and restaurants. . . Suh Jung Eui, a 10-year veteran at Hyundai, an economics graduate of Pusan University, now relegated to clerical work in the Hyundai group headquarters building was standing outside a bar. 'My seniors suggested I have a drink with them,' said Suh. And he was only a little surprised when he got to the rendezvous point to see men whom he did not recognize. . . .He was still more surprised when the thugs shoved fists and fingers into his back and elbows and ordered him into a waiting luxury car. . . . Suh said: 'They put me on the floor with their feet on top of me.' For five hours he remained in that position as the car plied the expressways to the southwestern port city of Mokpo. When he got to Mokpo, Suh found himself in a small room in a private house. 'They took off my blindfold and I ate a very little, just two spoonfuls.' His captors remained silent while he ate, after which one of them announced 'they would kill me with knives'." (Kirk, 1994)

Suh Jung-eui was held for five days and then released. Kirk wrote: "Suh woke up in the early morning of May 11. And discovering he was alone, Suh left the inn and hired a taxi to take him all the way to Seoul. . . The reason the kidnappers decided to let him go was clear as soon as he got there. 'The police were in front of my home. . . . They were looking for me.' All Suh knew for sure was that a member of the board of Hyundai Engineering and Construction . . . and the director of its general affairs department . . . were arrested for ordering the kidnapping, paying the six hoodlums . . . about $27,000, and later sentenced to a year apiece in jail. . . Suh looked back proudly on the lessons. He believed his experience had taught some of the top people at Hyundai: 'Management is trying not to have a union, but after this incident, their attitude changed a lot.'" (Kirk, 1994)

Strikes at Hyundai began. Kirk wrote: "No sooner had prosecutors begun interrogating Suh's kidnappers in Seoul than 3,000 workers struck against two plants of Hyundai Precision and Industry in Ulsan. The strike got rough when workers, on May 27, 1988, seized one of the factories, while the company chairman . . . was still at his desk. He was held (by the strikers) for five days along with ten of the top bosses before workers relented under threat of an attack by policemen outside the compound. Next door at Hyundai Motor, more than 20,000 workers walked out on May 30 demanding a 48% pay increase. Hyundai turned the walkout into a lockout, shutting its doors to everyone including those who wanted to report for work. . . . After 22 days, workers began showing up again at both HMC and Precision when HMC opened its doors. . . The union settled for a 28% raise but no pay for the days on strike. . . Hyundai put the cost of the strike at $490 million in lost sales of nearly 70,000 cars, 40,000 for export." (Kirk, 1994)

Hyundai was the major employer in Ulsan. Kirk wrote: "From Ulsan University, owned by Hyundai on the western fringes, to the Hyundai Department Store and the 'Ten Thousand Apartments' across from Hyundai Heavy Industries on the eastern peninsula, the reality was plain: Hyundai was Ulsan. Beyond the rice paddies and vegetable patches, Hyundai manufactures the motor vehicles, the railway containers, the pipes, the high-wire steel towers, the robots, the engines and heavy machinery, the deep-sea drilling rigs – plus the ships to carry all that Hyundai goods . . . to every seaport on earth." (Kirk, 1994)

The chaebols were constituted the economic heart of South Korea. The prosperity of Korean society depended upon how much wealth chaebols earned and how much they shared with the people who labored in producing the wealth. After conflicts, Hyundai management accepted independent labor unions.

KOREAN INDUSTRIAL STRUCTURE: MANAGEMENT AND LABOR

In the new Korean society, corruption by government officials was not as serious an economic problem as it had been in the old Joseon society. The new industrial problem was the (1) distribution of wealth between capital and labor and (2) employment opportunities, as shown in figure 13.2.

In the events of the industrial strikes in Ulsan, the *Action* consisted of a series of labor strikes between the *Groups* of management and labor in the *Process* of industrialization by Hyundai companies in Ulsan. The *Reason* of the strikes was wages, which the companies regarded only as costs, intending to reduce them to increase profits. The *System* was an industrial corporation, of which the Hyundai group had several businesses. The *Strategy* of the labor activists was to form a union and strike to increase wages. The *Policy* of the *Governance* of the Hyundai corporate boards was to not recognize unions, if they didn't have to recognize unions. This opposition resulted in a lapse in *Ethics*, when some Hyundai executives contracted for the kidnapping of a union activist. The *Performance* of the society was about the distribution of wealth in an *Infrastructure,* wherein Hyundai was the major employer in Ulsan. The *Ideology* of both groups of management and labor was capitalism. But the disagreement was over the *Principles* of rewards, when the wealth of the Hyundai group was growing at almost 30% annual rate, in which labor had only 3% wage increases. The resolution of the strikes led to compromises in rewarding labor, after the South Korean government legalized unionization acted in *Regulating* the Korean management-labor relations.

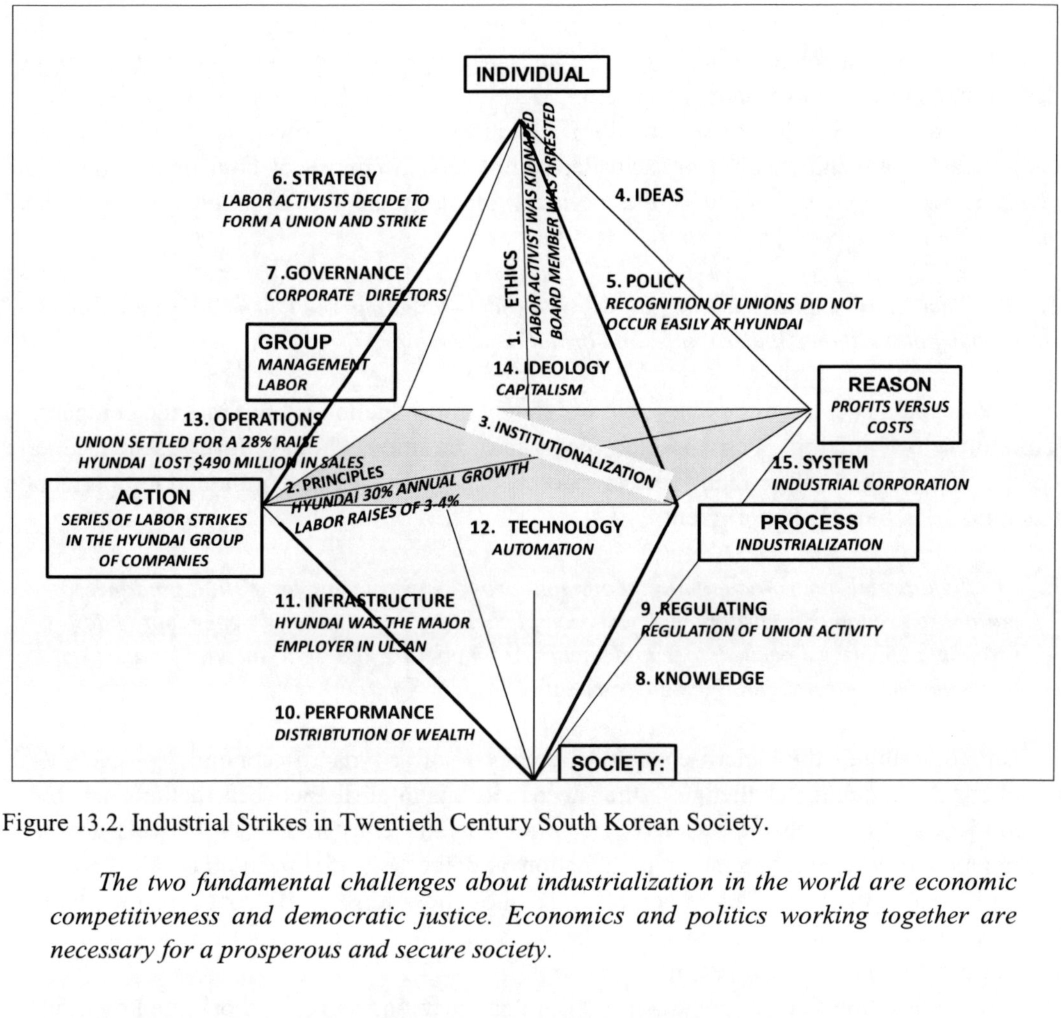

Figure 13.2. Industrial Strikes in Twentieth Century South Korean Society.

The two fundamental challenges about industrialization in the world are economic competitiveness and democratic justice. Economics and politics working together are necessary for a prosperous and secure society.

SELF-ORGANIZATION

A principal lesson we have learned in analyzing Korea's modern history is the importance of a proper industry-government partnership for economic development. And another lesson illustrated in this history is that both industry and government are self-organizing systems. A self-organizing societal system is one in which the control lies in local organization, not in a central authority. In a capitalistic democracy, even the government as a central authority, does not provide direct control over the political system or economic systems. (Here the self-organizing principle means that: local control in elections determines who leads the government for a time and local control in corporations determines the economic performance in a time.)

The economic system in South Korean society is a capitalistic system wherein control resides in the competition between economic agents over share of a free market (not a perfect economic market but a free economic market, meaning no monopolies in the market). The political system in South Korean society is a democratic system wherein control resides in free elections of government officials (not a perfect democracy but a free democracy, meaning elections are honest and representative). Thus competition is the key idea about control in a free-market economic system, and an honest representative election is the key idea about control in a just political system.

In contrast to South Korean freedom and justice, North Korea was organized as an ideological dictatorship, with a great leader completely in control of both its economic and governmental systems. This was the political ideology of the twentieth century called communism.

Communist/socialist ideologies were wrong – creating inefficient economies and starving populations, under unjust and brutal dictatorships.

For a capitalist economic system, the efficiency of production is built into the competitive marketplace of pricing. Perhaps, the economic tradition should have used the term ‘competitive market’ rather than ‘perfect market’ because it is competition and not perfection which makes a free market efficient.

Traditional macro-economic theory has proven insufficient for policy guidance of national economic development. A complementary contextually-dependent micro-economic theory is needed to supplement it – modeling the real societal contexts of particular nations, including its governments.

But describing the micro-economic contexts of a macro-economic system is a complicated modeling challenge. One needs to distinguish between actions in by a government and by a business – between politicians and entrepreneurs – between the self-organization of a political system of a nation and the self-organization of the economic system in global markets. This is complex. To show how this can be done, we next look in Figure 13.3, at a societal topology of explanations of both direct control of organizations and indirect control of self-organized infrastructures.

This depicts how self-organization in a capitalist democracy really works: (1) requiring a depiction of a business system focused upon markets and (2) a depiction of a political system focused upon a nation.

SELF-ORGANIZING CONTROL IN SOCIETY

The self-organizing view is seen by looking at the societal explanatory-relations from the bottom-up, from the society end of the ‘individual-society’ dichotomy. From this view, one sees that no single person is in charge of the basic societal systems of politics and of economy.

No one is in charge of a self-organizing society.

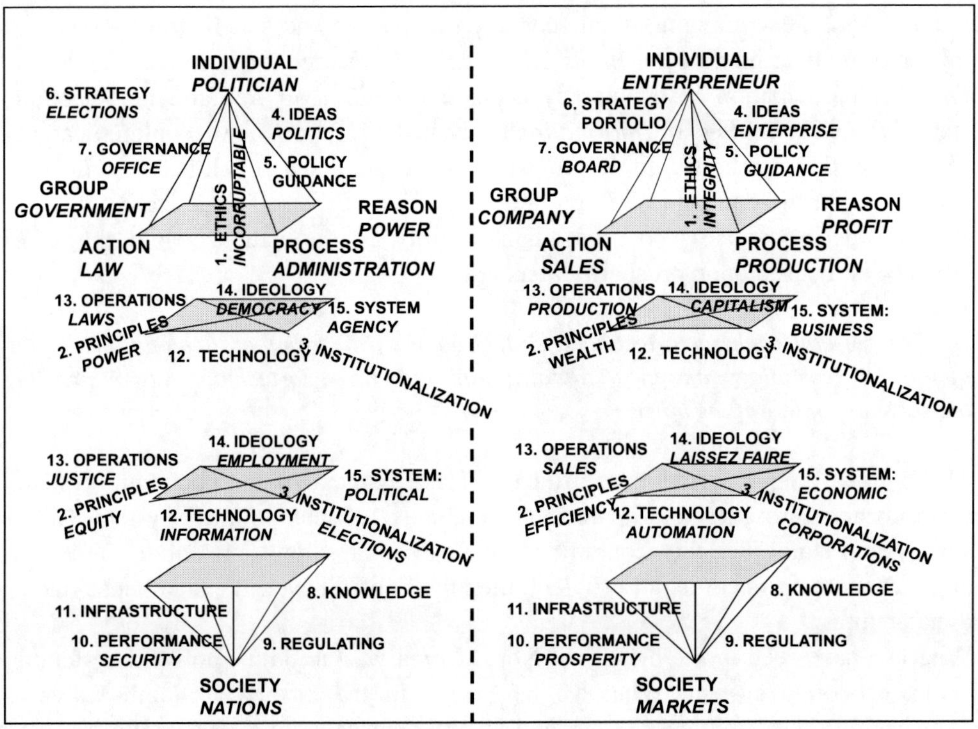

Figure 13.3. Government & Industry.

While a single business person (entrepreneur) can control a business system, no single entrepreneur is in control of the entire economic system. While a government official, such as a president or prime minister, can control the executive branch of government administration, no single politician is in control of the entire political system. This is why any capitalist democracy is self-organizing – not controlled by any central authority, business or government. No one person is in total control.

Then how can the socio-technical systems of self-organization work? How can the economic system work? How can the political system work? This has been the central problem of traditional economic theory; and this has been the central problem of traditional political theory.

Looking to the *right side* of the explanatory topology at the *Economic System*, one sees that the *Ideology* is *Laissez Faire* economic theory, which argues that business interests come first in society and all other interests in society, such as democracy and security, will follow. Within this *Ideology*, the *Principles* of the economic system focus upon efficiency, reducing costs of labor, and this is accomplished by improving the *Technology* of automation. The use of *Efficiency* is to increase *Sales* in markets from productive *Operations*. The *Institutionalization* of the economic system is through *Corporations* (legal economic agents with limitations of liability for owners).

In contrast, looking to the *left side* of the explanatory topology at the *Political System*, one sees that the ideology is *Employment*, with a goal of full employment as this provides economic security of all citizens. Within this ideology, the *Principles* of the political system focus upon *Equity*, encouraging all citizens to participate in and benefit from society. This is administered in government *Operations* providing equal *Justice* for all citizens under the rule of law. This participation in democracy is *Institutionalized* in free and fair elections and facilitated by the *Technology* of *Information* (universal education, news, communication).

This figure displays how complementary are the economic and political systems in a self-organizing capitalistic democracy. From the perspective of the self-organization of a society, the political performance is focused upon security in the nation, while the economic performance is focused upon prosperity in a nation.

We see that each socio-technical system has its own forms of ideology, principles, operations, technology, and institutionalization. And one socio-technical system provides the societal context of the other.

The political system provides a critical societal context of the economic system. The economic system provides a critical societal context of the political system. When the political system is not just, the economic system can collapse under rebellions and civil war. When the economic system is not efficient, then the political system can collapse under riots and dictatorship.

What we have seen in the history of North Korea was an unjust political system and an inefficient economic system. What we have seen in the history of South Korea is the development of a just political system and an efficient economic system. Politicians must primarily compromise, for a government to operate in a self-organizing democracy. Entrepreneurs must primarily compete, for businesses to operate in a self-organizing capitalistic economy.

Critical to the South Korean experience was a free political system and a free economic system and the proper cooperation between the two socio-technical systems. This is why studying the South Korean partnership between industry and government (between competitive entrepreneurs and incorruptible politicians) is important. The generalization from this study is basic to national economic development.

Government and industry must cooperate in a nation for a self-organizing democratic capitalist nation to operate effectively.

The partnership must be very ethical, with business people not corrupting politicians with bribes, and politicians not giving contracts to business people without fair and open competition.

'Laissez faire' economic ideology (which emphasized economics but demeaned politics) has been incomplete for effective economic development and democratic growth in an industrializing society.

A free market economy is necessary for economic growth, but there is a proper role of government in economic development; government needs to construct a proper context for economic activity – including an effective societal infrastructure, effective regulation of free markets (to insure safe products and an honest, stable financial system), and honest rule.

Direct Managed-Systems Control in Society

Individuals in a self-organized society have direct control over organizations, which can be seen as 'managed systems'. In industry, such organizations are businesses and corporations. In government, such organizations are government agencies, ministries, courts, parliaments. In Figure 14.3, one can see the individual-society dichotomy, from the individual's view, top-down over the explanatory topology. The two different kinds of individuals, entrepreneurs and politicians, think differently.

Entrepreneurs have *Ideas* about business as a kind of *Enterprise*, as system of producing goods and services and selling them. Their *Strategy* focuses upon a *Portfolio* of businesses in which to engage. Their *Policies* formulate *Guidance of the Processes of* a given business. *Governance* consists of selecting a *Board* of Directors for the corporation. The *Group* which they directly control is a *Company*. The *Action* for success of a company lies in *Sales*. And the *Reason* underlying all the business activity is *Profit*. In *Ethics*, entrepreneurs should, for the good of markets, act with *Integrity*.

In contrast, *Politicians* have *Ideas* about government as *Politics*, as a system of gaining power and exercising power. Their *Strategy* focuses upon winning Elections to hold power. Their *Policies* formulate *Guidance for* the directions of government activities. *Governance* consists of holding an elective or appointive *Office*. The *Group* which they directly control consists of government agencies. The *Action* for success of a company consists of passing and enforcing *Laws*. And the *Reason* underlying all the political activity is *Power*. In *Ethics*, a politician for the good of the nation, acts *Incorruptibly* (without corruption).

Comparing these individuals of business and government, one sees how differently they think and how different are the contexts of their actions! Getting them to cooperate, without improper interference in their different domains, is a difficult and delicate task. In South Korea for its successful economic growth, President Park succeeding in establishing proper cooperation with chaebols founders, such as Chairman Lee of the Samsung Group and Chairman Chung of the Hyundai Group. Later presidents, beginning with Roh Tae-woo and Kim Young-sam, succeeded in establishing democracy in South Korea.

If politicians are corrupt and entrepreneurs lack integrity, a proper national partnership between government and industry cannot be attained. Corruption by political leaders and fraud by business leaders are the two major factors of why self-organizing societies do not develop economic strength and political justice.

Limiting Abuses of Leadership in Political and Economic Systems

How can a self-organizing society limit abuses by individuals in leadership positions in government and in industry, in the self-organizing political and economic systems of the society? To answer this, we turn to Figure 13.4, to focus on the bottom half of the explanation topology of a self-regulating society - particularly upon the explanatory relationships of *(2) Principles, (3) Institutionalization, (9) Regulating.* These can have a major impact upon the ethical contexts of the actions of individual within a self-organizing society. These can have important influences upon limiting or permitting ethical abuses of power by leading individuals (who control the managed-systems, organizations, of government and industry).

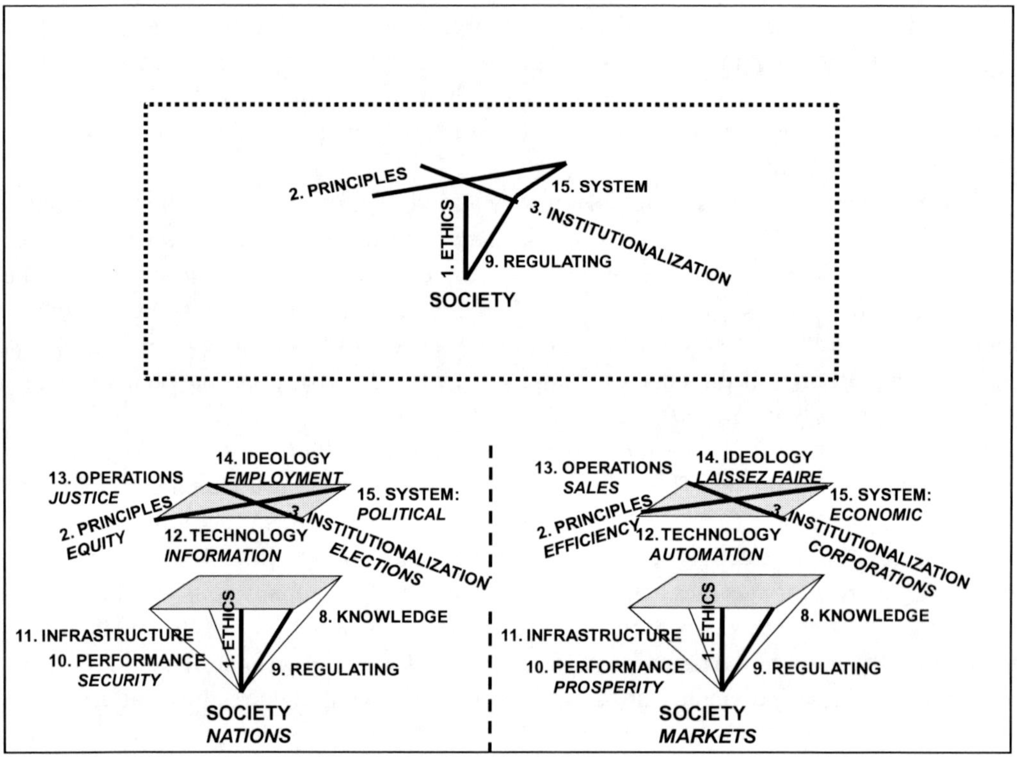

Figure 13.4. Self-Organizing Relationships Which Limit Individual Corruption and Fraud.

*Principles (*in the self-organizing socio-technical systems of politics or economics) have an influence over how individuals (politicians or entrepreneurs) can act in managing their own governmental or industrial organizations. For example in a dictatorship, corruption by government officials is permitted in principle. In the German Nazi government of the twentieth century, the looting of Jewish citizens properties was permitted in principle by Nazi Regime, and so the Nazi officials looted Jewish homes, businesses, and banks. The abuse of taxation was permitted by government officials in the late Joseon dynasty.

Institutionalization (in the self-organizing socio-technical system of politics or economics) has an influence over how individuals (politicians or entrepreneurs) can act in managing their own governmental or industrial organizations. For example in U.S financial industry in 1999, banking integration of investment and commercial banking was enacted into law. Commercial bank executives, seeking the large annual bonuses of investment bankers, integrated commercial and investment banks – which created the 'too-big-to-fail banks' that failed in 2008 and had to be bailed out by the U.S. government.

Regulating (in the self-organizing socio-technical system of politics or economics) has an influence over how individuals (politicians or entrepreneurs) can act in managing their own governmental or industrial organizations. For example in the U.S. financial industry in 2000, government officials argued against the regulation of the new financial derivatives market, and it in 2008 (with fraudulently concocted and rated mortgage derivatives) the derivatives collapsed the global financial system.

In a self-regulating society, the ethics of individuals acting in directly-managed organization are constrained (or not constrained) by the proper (or improper) Principles, Institutionalization, and Regulating aspects of the socio-technical systems of a society.

Without proper constraint, greedy individuals (by irresponsibility, fraud, or corruption) can abuse the institutions of a self-organizing society.

The history of South Korea has shown remarkable progress by remarkable individuals (in government and in industry) toward establishing in the self-organizing systems of Korea, their proper Principles, Institutionalization, and Regulating of the political and economic systems – toward a just democracy and a competitive economy.

CONCLUSION

What we have seen in our analysis of the development of South Korea is empirical evidence for the following postulates.

1. Democratic and capitalistic societies provide prosperity and justice for their populations; instead dictatorships and communistic societies yield poverty and injustice for their populations.
2. Traditional economic theory is a correct concept for an efficient economic system, provided that markets therein operate freely. However, free markets do not occur independently, without proper government regulation for safety and integrity. This is to say, economic theory is not a 'context-less' theory. Economic theory is not always empirically true, independent of societal context. Economic theory applied to any real society is dependent upon the contexts of the other systems of a society – political, cultural, and science/technology contexts.
3. As an ideology, Laissez faire economic theory assumed that a free-market economy will automatically create a democracy, with justice, prosperity, knowledge, security. It is a 'utopian' vision of what an industrial society might become. But historically, no real society has functioned effectively as a 'let-be' economy, without proper government regulation. The desirable societal features (such as prosperity, justice, knowledge, security) must all be constructed within a proper partnership between private industry and government.
4. If a nation were isolated, then constructing a partnership between private industry and government might be easy. However, a proper partnership must be constructed in a global context. Therein industry must be capable of exporting internationally and competitively; and a given nation's financial system will be partly dependent upon international capital flows.
5. In international finance, policies (to apply Laissez-faire economic ideology by international financial organizations) was not effective. As evidenced in the Asian

Financial crisis, the Laissez-faire policies of the IMF didn't work, but only accelerated and deepened those national depressions.

6. Abuse-of-power by government leaders and/or excessive-greed by business leaders are the two principle flaws causing self-organizing to societies fail – why democracies fail and why financial systems fail.
7. The ethics of individuals are constrained (or not-constrained) by the proper (or improper) *Principles, Institutionalization, and Regulating* aspects in the socio-technical systems of a self-organizing society.

REFERENCES

Abernathy, William J. 1978. *The Productivity Dilemma*. Baltimore, Md. USA: Johns Hopkins University Press.

Abernathy, William J. and K. B. Clark. 1985. "Mapping the Winds of Creative Destruction", *Research Policy*, vol. 14, no. 1.

Amnesty International. 2011. North Korea: Political Prison Camps. May 3. (http://www.amnesty.org/en/library/asset/ASA24/001/2011/en/2671e54f-1cd1-46c1-96f1-6a463efa6f65/asa240012011en.pdf)

Ayres, Robert U. 1989. "The Future of Technological Forecasting", Technology Forecasting and Social Change, Vol. 36, No. 1-2, August, pp. 49-60.Bang, Jung-hyun. 2010. Hail the Father of Business, Lee.

Betz, Frederick. 1987. *Managing Technological Innovation*. Prentice Hall.

Betz, Frederick. 1998. *Managing Technological Innovation.* John Wiley & Sons.

Betz, Frederick. 2011. *Societal Dynamics.* New York: Springer.

Blustein, Paul. 2003. *The Chastening: Inside the Crisis That Rocked The Global Financial System And Humbled The IMF.* New York: Public Affairs Books

Blustein, Paul. 2005. And the Money Kept Rolling In (And Out), New York: Public Affairs, Perseus Books Group.

Boettke, Peter J. 1997. "What Went Wrong with Economics?", Critical Review Vol. 11, No. 1, pp. 35-58. (http://www.the-dissident.com/Boettke_CR.pdf)

Brunnel, David, 2000. *Making the Cisco Connection*. John Wiley & Sons.

Byung-chul, Korea IT Times, February 11. (http://www.koreaittimes.com/story/7223/hail-father-business-lee-byung-chul)

Carayannis, E.G. and D.F.J. Campbell. 2009. "'Mode 3' and Quadruple Helix: toward a 21st century radical innovation ecosystem", *International Journal of Technology Management* 46 (3-4).

Carayannis, E.G. and D.F.J. Campbell. 2010. "Triple Helix, Quadruple Helix and Quintile Helix and How Do Knowledge, Innovation, and Environment Relate to Each Other?" *International Journal of Technology Management* 1 (41-69).

Carayannis, E. G. and D.F.J. Campbell. 2012. *Mode 3 Knowledge Production in Quadruple Helix Innovation Systems*, Springer Briefs in Business 7.

Chandler, Alfred. 1990. "The Enduring Logic of Industrial Success," *Harvard Business Review*, March-Aril, pp. 130-140.

Choi, Youngrak. 1996. *DynamicTechno-Management Capability*. Avebury.

Choi, Youngrak. 2007. "Growth Path of Samsung Semiconductor", *Innovation Studies*, Vol 2, No. 1, June.

Choi, Youngrak. 2010. "Korean Innovation Model, Revisited". *Science Technology and Innovation Policy Review*. Vol 1 No. 1, Spring.

Clifford, Stephanie. 2013. "US Textile Plants Return, With Floors Largely Empty of People", *New York Times*, September 19.

Coe, David and Se Jik Kim. 2002. *Korean Crisis and Recovery*, International Monetary Fund and Korea Institute for International Economic Policy.

Choe, Sang Hun. 2013. "South Korea Pushes Back on North's Threats", *The New York Times*. March 6.

Choe ,Sang Hun, 2013a. "Prosecutors Raid Home of Former South Korean President", New York Times, July 16.

Choi, Youngrak. 2010. "Korean Innovation Model, Revisited". Science Technology Innovation Policy Review. Vol 1, No. 1, Spring.

Chopra, Ajai, Kenneth Kang, Meral karasulu, Hong Liang, Henry Ma, Anthony Richards. 2002. "From Crisis to Recovery in Korea: Strategy Acheivements, and Lessons". (Coe, David T. and Se-jik Kim. 2002. Korean Crisis and Recovery. International Monetary Fund: Korea Institute for International Economic Policy.)

Connaughton, R. M. 1988. The War of the Rising Sun and the Tumbling Bear—A Military History of the Russo-Japanese War 1904–5. London. ISBN 0-415-00906-5.

Currie, Antony, Rob Cox and Martin Hutchinson. 2009. "GM Bankruptcy Loomed for Years", *New York Times,* June 2.

Eckert, Carter J. 1991. Offspring of Empire, The Koch'ang Kims and the Colonial Origins of Korean Capitalism, 1876-1945. University of Washington Press

Ford, David and Chris Ryan. 1981. "Taking Technology to Market," *Harvard Business Review,* March-April.

Freeman, Chris., 1994. "The Economics of Technical Change: A Critical Survey Article". *Cambridge Journal of Economics,* 18 (5).

Glaeser, Edward L. 2009. New York Times, http://economix.blogs.nytimes.com/2009/10/06/what-happened-to-argentina/

Goodman, Peter S. 2008. "Taking Hard New Look at a Greenspan Legacy" New York Times, October 8.

Günterberg, B. and G. Kayser. 2004. "SMEs in Germany - Facts and Figures 2004". IfM-Materialien Nr. 161. Bonn: Institut für Mittelstandsforschung. Retrieved 2012-09-26.

Krugman., Paul. 2009. "The B Word". The New York Times, March 17.

Harden, Blaine. 2009. "North Korea's Hard-Labor Camps: On the Diplomatic Back Burner", Washington Post Foreign Service, July 20. (http://www.washingtonpost.com/wp-dyn/content/article/2009/07/19/AR2009071902178_pf.html

HEC. 2013. http://hyundaiengineering.blogspot.kr/2010/11/hyundai-conglomerate-groups-of-korea_7719.html

Howells, J. 2006. "Intermediation and the role of intermediaries in innovation," *Research Policy,* Vol. 35, no. 5, pp. 715–728, June.

Hyundai. 2013. http://worldwide.hyundai.com/wcm/idc/groups/sggeneralcontent/@hmc/documents/sitecontent/mdaw/mdm5/~edisp/hw039016.pdf

Innace, Jospeph J. and Abby Dress. 1992. *Igniting Steel: Korea's POSCO Lights the Way*. New York: Global Village Press.

Jetter, Antonie J. 2011. "The 'Fraunhofer Institute for Production Technology': An example of multiple roles of government". Guest Lecture at Willamette University

Jetter, Antonie J. and Songphon Munkongsujarit. 2013. "The Impact of Social Capital on the Outcomes of Innovation Intermediation Process" PICMET 2013 Conference, San Jose, California. USA.

Kang, Myung Hun. 1996. The Korean Business Conglomerate: Chaebol Then and Now, University of California Berkely, Center for Korean Studies.

Karl Polanyi Institute of Political Economy. 2013. Concordia University, 1455 de Maisonneuve West; Montreal, Quebec, Canada H3G 1M8. (http://polanyi.concordia.ca//about)

Kim, Byung Kook, Chang Jae Baik, Ezra F. Vogel, and Jorge I. Domínguez. 2011. The Park Chung Hee Era: The Transformation of South Korea. Harvard University Press.

Kim, Dae Il. 1997. "Demand for College Education and the Labor Market Outcomes in Korea." APEC-HRK-NEDM-KDI International Seminar on Improving the Economic Performance of Education, November 5-9. Seoul: Korea Development Institute.

Kim, Hyung A. 2012. Korea's Development Under Park Chung-Hee, Routledge/Asian Studies Association of Australia (ASAA) East Asian Series), ISBN-10: 0415323290 (First published, 2004 by Routledge, London)

Kim, Woo Choong. 1992. *Every Street Is Paved with Gold*. New York: William Morrow and Company.

Kirk, Donald. 1994. *Korea Crisis*. New York and London: St. Martins Press.

Kirk, Donald. 1994a. *Korean Dynasty: Hyndai and Chung Ju Yung*. Asia 2000 Ltd.

Kruger, Anne O. and Yung Ho Yoo. 2002. "Falling Profitability, Higher Borrowing Costs, and Chaebol Finances During the Korean Crisis". (In David Coe and Sejik Kim, *Korean Crisis and Recovery*, IMF, 2002)

Krugman, Paul. 1994. "The Myth of Asia's Miracle". Foreign Affairs 73 (6): 62–78. JSTOR 20046929.

Landler, Mark. 2013. "Detecting Shift, U.S. Makes Case to China on North Korea", New York Times. April 5.

Lee, Chong Sik. 2012. *Park Chung-Hee: From Poverty to Power*. Palos Verdes, CA: KHU Press.

Lee, Kyu Uck. 1986. "The Concentration of Economic Power in Korea", *Industrial Development Policies and Issues.*

Lee, Kyu Sung, 2011. *The Korean Financial Crisis of 1997*. The World Bank and Korean Development Institute. B004U6WN2I.

Letzkowitz, H. and L. Leydesdorff. 2000. "The Dynamics of Innovation: from National Systems and 'mode 2' to a triiple Helix of University-Industry-Government Relations." *Research Policy*, 29, (2).

Lowenstein, Roger. 2008. "Siphoning G.M.'s Future", *New York Times*, July 10.

Mako, William P. 2002. "Corporate Restructuring and Reform: Lessons from Korea". (In David Coe and Sejik Kim, *Korean Crisis and Recovery*, IMF, 2002)

Maynard, Micheline. 2008. "At G.M., Innovation Sacrificed to Profits", *New York Times*. December 5.

Maynard, Micheline. 2009a. "The Steady Optimist Who Oversaw G.M.'s Decline", *New York Times*, March 29.

McAfee, R. Preston, Hugo M. Mialon, and Michael A. Williams. 2003. "Economic and Antitrust Barriers to Entry", (http://www.mcafee.cc/Papers/PDF/Barriers2Entry.pdf)

Melamed, Claire,Renate Hartwig and Ursula Grant 2011. "Jobs, growth and poverty: what do we know, what don't we know, what should we know?" London: Overseas Development Institute, (http://www.odi.org.uk/sites/odi.org.uk/files/odi-assets/publications-opinion-files/7121.pdf)

Meyer-Stamer, Jorg and Frank Wältring: 2000. "Behind the Myth of the Mittelstand Economy, The Institutional Environment Supporting SME Enterprises in Germany", Institute for Development and Peace at the Gerhard-Mercator-University Duisburg. (http://www.mesopartner.de/fileadmin/user_files/working_papers/INEF_report46_at_www.mesopartner.com.pdf)

Mitchell, Tony. 2010. *Samsung Electronics*. John Wiley & Sons.

More, Gordon E. 1965. "Cramming More Components onto Integrated Circuits", *Electronics* Magazine, Vol. 38, No. 8, April 19. (http://download.intel.com/museum/Moores_Law/Articles_Press_Releases/Gordon_Moore_1965_Article.pdf)

Mount, Ian. 2011. "Argentina's Turnaround Tango", New York Times. September 1, (http://www.nytimes.com/2011/09/02/opinion/argentinas-turnaround-tango.html)

Oberdorfer, Don. 2001. *The Two Koreas*. Basic Books.

OECD. 1997. *National Innovation Systems*. Organization For Economic Co-Operation And Development. (http://www.oecd.org/science/innovationinsciencetechnologyandindustry/2101733.pdf)

OECD. 2006. *National Innovations Systems: Switzerland*. Organization for Economic Development Co-operation and Development

O Won Chul. 2009. *The Korea Story*. Korea: Wisdom Tree Publishing (www.wisdomtree.co.kr)

Park, Chung Hee. 1962. *The Country, The Revolution, and I*. Korea: Hollym Corporation. (first edition 1962, second edition 1970)

Park, Kisung. 2008. "Military Authoritarian Regimes and Economic Development", Thesis, Naval Postgraduate School, Monterey, Califonia, USA.

Polak, Jacques J. 1998. "The IMF Monetary Model at 40", in *Economic Modeling,* (Elsevier Science) (Vol. 15 (July), pp. 395-410., July 1998.

Polak, Jacques J. 1989. *Financial Policies and Development.* Organization for Economic Co-operation and Development.

Polak, Jacques J. 2005. *Economic Theory and Financial Policy*. Edited by James M. Boughton. International Monetary Fund. London: M. E. Sharpe.

Polyani, Karl. 1944. *The Great Transformation.* Beacon Press (reprinted in 2001)

Porter, Michael, 1985. *Competitive Strategy*, New York: The Free Press

Power, John and Choi In Jeong. 2013. "How should small business be supported?", *Korean Herald*, April 8.

Quinn, James Brian,1988. "Strategies for Change", (in *The Strategy Process,* ed by James B. Quinn, Henry Mintizberg and Robert M. James, Prentice Hall), Englewood Cliffs, New Jersey, pp 2-8.

Qiu, Yuting and Yiyang Fan. 2013. "Rethinking Global Innovation Strategy: Emerging Market Perspectives". *Business and Management Research*. Vol. 2, No. 3.

Ray, George F., 1980. "Innovation and the Long Cycle", in B. A.,Vedin (ed.), *Current Innovation*, Almqavist & Wiksell, Stockholm.

Reid, T. T. 1985. "The Chip", *Science 85*, February pp 32-41.

Reinhart, Carmen M. and Kenneth S. Rogoff. 2009. *This Time is Different Princeton.* New Jersey: University Press.

Reinhart, Carmen and Graciela L Kaminsky. 2001. "Bank Lending and Contagion: Evidence from the Asian Crisis". In *Regional and Global Capital Flows: Macroeconomics Causes and Consequences, NBER-EASE*, Volume 10, Editors: Takatoshi Ito and Anne O. Krueger, University of Chicago Press, 2001, ISBN: 0-226-38676-7 (http://www.nber.org/chapters/c10732.pdf)

Robaschik, Frank. 2013 "Medium-sized Business in Germany: Ecoomic Power and Success Factors of Mittelstand". Korean-German Seminar on the German Mittelstand. Korea Federation of Small and Medium Business and Embassy of the Federal Republic of German and Germany Trade & Investment. May 3, Lilly Hall, Seoul, Korea.

Roubini, Nouriel and Stepen Mihm. 2010. *Crisis Economics.* New York: Penguin Press.

Samsung Foundation, 2012. *Biography of Lee Byung-Chull.* (http://hoamprize.samsungfundation.org)

Sanger, David, Jeff Zeleny and Bill Vlasic. 2009. "G.M. to Seek Bankruptcy and a New Start", *New York Times*, June 1.

Santayana, George. 1905. "*The Life of Reason*, Volume 1'. Scribner's Sons, (republished 1998 by Prometheus Books)

Schumpeter, Joseph. 1942. *Capitalism, Socialism, and Democracy*. Harper & Brothers (Reprinted in 2008, Harper Perennnial Modern Classics)

Shin, Soonchul and Jinyoung Lee. 2008. *A Short History of the Donghak Peasant Revolution*. Donghak Peaant Revolution Memorial Association. (English edition: translated by Rohini Sing and Chongmin Lee.

Siber, Sandra and Jospe Valor. 2007. "Google: Organizing Information on the Web", Case Study,niversity of Navarra, IESE Business School.

Simon, Hermann. 1996. *Hidden Champions*. Boston: Harvard Business School Press.

Soros, George. 2004. *The Crisis of Global Capitalism*. ISBN-10: 8176497991

Southerton, Donald. 2012. *Hyundai and Kia Motors The Early Years and Product Development*. Amazon Kindle Edition.

Stanfield, J. Ron. 1986. *The Economic Thought of Karl Polanyi: Lives and Livelihood*. Macmillan, ISBN 0-333-39629-4

Steele, Lowell W. 1989. *Managing Technology: The Strategic View,* McGraw-Hill, New York.

Steers, Richard M. 1999. *Made in Korea: Chung Ju Yung and the Rise of Hyundai.* Routledge.

Stevenson, Howard and David Gumpert. 1985. "Heart of Entrepreneurship', *Harvard Business Review*, March 1.

Stiglitz, Jospeth E. 2002. *Globalization and Its Discontents.* W. W. Norton and Company

Steil, Benn. 2013. *The Battle of Bretton Woods*. Princeton University Press.

Suh, Joong Hae and Derek H.C. Chen (editors). 2007. *Korea as a Knowledge Economy.* Korea Development Institute and The World Bank Institute. Washington D.C., The World Bank.

Sunoo, Harold Hakwon. 1994. *20th Century Korea*. Seoul, Korea: Nanam Publishing House.

Uchida, Jun. 2011. Brokers of Empire: *Japanese Settler Colonialism in Korea, 1876–1945.* http://www.wilsoncenter.org/modern-korean-history-portal

Utterback, James M. 1996 *Mastering the Dynamics of Innovation*. Harvard Business School Press. 1996.

Venohr, Bernd and Klaus E. Meyer. 2007. "The German Miracle Keeps Running: How Germany's Hidden Champions Stay Ahead in the Global Economy", *Working Paper* 30, FHW Berlin.(http://www.berndvenohr.de/download/veroeffentlichungen/forschung/The_German_Miracle_Keeps_Running.pdf)

Venohr, Bernd. 2010. "The Power of Uncommon Sense Management Principles – The Secret Recipe of German Mittelstand Companies", 2nd Global Drucker Forum Venia 2010, November 18 and 19. (http://www.druckersociety.at/repository/2010/day01/15%2730-17%2700/Venohr_101118_PPT_Beamerversion.pdf)

Vlasic, Bill. 2008. "G.M., Once a Powerhouse, Pleads for Bailout", *New York Times*, November 11.

Wolff, Michael F. 1983. "Mervin J. Kelly: Manger and Motivator", *IEE Spectrum,* May, pp. 81-85.

Woods, Ngaire. 2006. *The Globalizers: The IMF, the World Bank, and Their Borrowers*. Cornell University Press.

World Steel Association. 2012. http://www.worldsteel.org/statistics/top-producers.html

Yonhap News. 1999. "Demise of a great name." *Asian Pulse*. August 19. (http://www.atimes.com/atimes/Korea)

Yonhap News. 2005." Daewoo founder sentenced to 10 years in prison", *Asian Times* http://www.atimes.com/atimes/Korea/HF01Dg03.html"

Yonhap News. 2013. "North Korea nuclear arms program now a 'real threat,' Seoul says". *The Korea Herald*. Sept 4.

Yoon, Je Cho. 2002. "What Have We learned from the Korean Economic Adjustment Program?" (Coe, David T. and Se-jik Kim. 2002. Korean Crisis and Recovery. International Monetary Fund: Korea Institute for International Economic Policy.

Ypsilanti, Dimitri. 1985. "The Semiconductor Industry", *The OECD Observer*, No 132.

INDEX

D

E

F

G

H

I

J

K

Q

R

S

T

U

V

W

Y

Z